Petroleum
and Structural Change
in a Developing Country:

The Case of Nigeria

Peter O. Olayiwola

PRAEGER

New York
Westport, Connecticut
London

Library of Congress Cataloging-in-Publication Data

Olayiwola, Peter O.
 Petroleum and structural change in a developing
country.

 Bibliography: p.
 Includes index.
 1. Nigeria – Economic conditions – To 1960.
2. Nigeria – Economic conditions – 1960–
3. Nigeria – Economic policy. 4. Petroleum industry
and trade – Nigeria. I. Title.
HC1055.043 1986   330.9669′05   86-21216
ISBN 0-275-92115-8 (alk. paper)

Library of Congress Catalog Card Number: 86-21216
ISBN: 0-275-92115-8

First published in 1987

Praeger Publishers, 521 Fifth Avenue, New York, NY 10175
A division of Greenwood Press, Inc.

Printed in the United States of America

The paper used in this book complies with the Permanent
Paper Standard issued by the National Information Standards
Organization (Z39.48-1984).

10 9 8 7 6 5 4 3 2 1

To my parents,
Elisha O. and Abigail F. Olayiwola

Preface

In trying to determine the basic problem that has caused Nigeria's growth without change, my own thoughts about Nigeria's major problem have undergone a transformation. I had previously believed that the basic problem was a lack of efficient and effective leadership, that Nigeria could have broken the underdevelopment syndrome if it had had competent leaders running the government. There were several reasons for that belief. Born and raised in Nigeria, I witnessed the transformation of its economy from an export-agriculture one to a petroleum-based one. I was in Nigeria when it gained political independence from Britain in 1960, during the civil war, and during the oil boom. Throughout those periods, the direction of the country seldomly seemed sure; the leadership was often challenged.

During my research, I came to realize that even with the right people in the right positions, Nigeria's condition would have been the same. This is because, as I discovered, the lack of autonomous development in Nigeria can be traced to a structural condition of underdevelopment.

The situation in Nigeria is a complex one. Hence, this study is an attempt to provide a comprehensive analysis of the political economy of Nigeria from 1860 to 1985. This allows the book to cover all aspects of national development efforts in Nigeria before and after independence. Nigeria's failure to achieve autonomous development despite the oil wealth of the 1970s is only a symptom of a specific problem. This book offers a solution to that puzzle.

This study is intended to provide a basis for further research into the problem of autonomous development and its attainment in Nigeria, and also in other developing countries that are facing problems similar to those confronting Nigeria.

Acknowledgments

In this short space, it is impossible to acknowledge everyone who, directly and indirectly, has contributed to the completion of this study. Nevertheless, let me single out a few.

My special gratitude goes to Prof. John Byrne, director of energy and urban policy research at the College of Urban Affairs and Public Policy, University of Delaware. I deeply appreciate his guidance, understanding, and encouragement, as well as his sincere interest in this study, as evidenced by his thorough editing of the chapters and by his constructive comments. My gratitude also goes to Profs. William W. Boyer, Olasupo Ladipo, Dan Rich, and Robert Warren. I appreciate their valuable and constructive comments.

Many thanks to my wife, Olayinka, and to my children; the many sacrifices they made are greatly appreciated. I am also most grateful to my editor, Ms. Catherine Woods, for the professional way in which this project was handled and for her encouragement.

Of course, none of the persons mentioned above should be held responsible for views expressed in this study. The responsibility for any weakness, error, or omission is entirely mine.

Contents

Tables

Figures

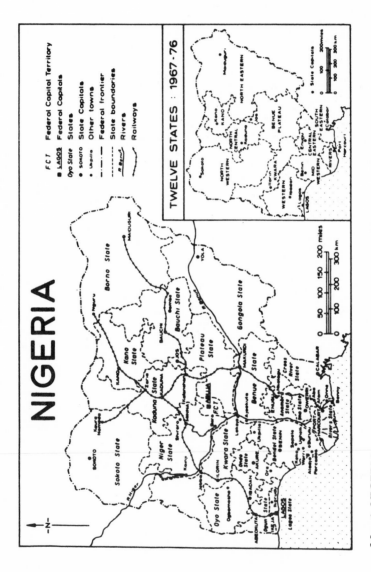

Map 1. NIGERIA

Source: Anthony Kirk-Greene and Douglas Rimmer, *Nigeria Since 1970*
(London: Hodder and Stoughton, 1981), p. xiv. Reprinted by permission.

1 Growth Without Change: The Case of Nigeria

INTRODUCTION

This book examines the development experience of Nigeria since political independence in 1960. As a society, Nigeria has undergone profound changes. It was transformed from a primarily agricultural society into an industrializing one. A key source of this change has been the emergence of the petroleum economy.

As calculated from the International Financial Statistics (IMF 1984:454–55), Nigeria's gross domestic product (GDP) grew an extraordinary 81 percent per annum, on average, between 1960 and 1980. The aggregate performance of the petroleum economy was far better, however, adding to Nigerian economic wealth at an average rate of 7,400 percent per annum. Nigeria's accumulated petroleum wealth from independence to 1980 was ₦ 58.83 billion, compared with accumulated GDP of ₦ 267.25 billion, indicating that the petroleum economy accounted for 22 percent of aggregate economic growth. Distributed across the Nigerian population (which grew by 33 million, or 60 percent), petroleum earnings meant an addition to annual per capita income of ₦181.41, while total per capita income grew by ₦ 505.63 during the same period. Thus, the petroleum economy, without consideration of multiplier effects, was responsible for 36 percent of all per capita economic growth. Judged on these terms, Nigeria's overall economic performance from independence to 1980 was spectacular; the performance of its petroleum economy was astounding.

Despite this performance, the structure of Nigeria's political economy is nearly the same as it was at independence. It remains one in which economic life depends critically upon world market conditions and the level of trade with developed economies, just as colonial Nigeria did. This dependence has been painfully clear to Nigeria. Between 1980 and

1983, petroleum export revenues fell from ₦13.999 billion to ₦7.786 billion, representing a decline of ₦6.213 billion, over 44 percent. In addition, little or no political development has taken place despite political independence. The government is still non democratic. In describing several former British colonies, including Nigeria, Ferrel Heady (1979:314–15) argued that following political independence, "continuity has proved to be greater in the administrative than in the political sphere, with bureaucratic elites formed during the colonial period gradually taking over political control in what were immediately after independence polyarchal competitive political regimes." The most common direction of change in developing countries such as Nigeria has been toward bureaucratic elitism. Nigeria's military supplanted the short-lived competitive parliamentary government installed by Britain at independence and the polyarchal competitive presidential form of government that lasted from 1979 to 1983.

TABLE 1.1.   Social Indicators, 1960–80

Indicators	1960	1965	1970	1975	1980
Population					
Total (in millions)	51.6	58.5	66.2	74.9	84.7
Urban population (% of total)	13.1	14.7	16.4	18.2	20.4
Demographic characteristics					
Life expectancy (years)	38.7	41.2	43.7	46.2	48.6
Health and nutrition					
Population per physician	73,710	44,990	24,670	17,630	12,550*
Pop. per nursing person	4,040	6,250	5,070	3,820	3,010*
Pop. per hospital bed	3,030	2,430	2,220	1,390	—
Calorie supply per Capita					
(% of requirements)	83.5	—	82.1	80.1	91.0
Infant mortality rate					
(per thousand)	183.4	170.4	158.0	146.3	135.2
Education					
Primary school enrollment ratio	36.0	32.0	37.0	53.0	98.0*
Secondary sch. enrollment ratio	4.0	5.0	4.0	8.0	16.0*
Adult literacy rate	15.4	—	—	—	34.0
Employment (%)					
Labor force participation rate	42.2	40.8	39.3	37.9	36.4
Labor force in agriculture	71.0	66.7	62.0	58.1	54.0
Labor force in industry	10.0	11.9	14.0	16.4	19.0

*1979.

Source: World Bank, 1983b: II, 69.

Not only has Nigeria shown little economic and political progress since independence, but its social development has been quite slow in several dimensions. Table 1.1 summarizes selected social indicators for 1960 to 1980. During the 20 years following independence, the (estimated) population grew from 51.6 million to 84.7 million and the urban population as a percent of total population increased from 13.1 to 20.4. Life expectancy showed some improvement; it increased from 38.7 to 48.6 years. Access to health care, as measured by population per physician, population per nursing person, and population per hospital bed, showed remarkable improvement. However, what these statistics fail to disclose is the decline in the quality of health care delivery, such as frequent shortages of drugs, hospital equipment and supplies, and the deteriorating level of morale among hospital employees. Access to education, as measured by primary school enrollment ratio, improved dramatically from 36 percent in 1960 to 98 percent in 1979. The universal primary education program played a major role in this growth. Nigerian population was better educated in the 1970s than in the 1960s, and despite the oil boom of the 1970s, the labor force participation rate declined from 42.2 percent in 1960 to 36.4 percent in 1980. In part, the decline was due to the failure of the oil boom to generate enough employment opportunities to keep pace with population growth. But equally important, the percentage of the labor force in agriculture declined from 71 percent in 1960 to 54 percent in 1980—a direct result of an educational system that prepared students to look forward to white-collar jobs and deemphasized farming.

When compared with the social indicators for the republic of Cameroon and Niger (its neighbors), Nigeria, despite the oil boom of the 1970s, showed little relative progress (see Table 1.2). While Nigeria and Cameroon are nearly identical in GNP per capita (in 1981, Nigeria posted a GNP per capita of U.S.$870, while Cameroon's stood at U.S.$880), Cameroon's educational and health care performance has been superior. Moreover, Cameroon has been able to maintain its agricultural system, and as a result has experienced a less severe drain on foreign exchange resources due to agricultural imports (compare GDP sectoral figures for agriculture in World Bank 1983 b:I). Niger, which is a much poorer country than Nigeria (GNP per capita of U.S. $330 in 1981), has nevertheless shown nearly equal progess on certain health care indicators (population per hospital bed and calorie supply per capita) despite a severe drought during the 1970s. Its labor force participation rate was not much below that of Nigeria, although its per capita GNP was less than half of Nigeria's. In sum, Nigeria's progress in several social development areas was well within the range experienced during 1960 to 1980 by its neighbors in West Africa.

The central argument of this study is that Nigeria has experienced growth without change, and as a result has failed to establish the basis for

TABLE 1.2. Cameroon and Niger: Social Indicators, 1960 and 1980

Indicator	Cameroon		Niger	
	1960	1980	1960	1980
Population				
Total (in millions)	5.6	8.4	2.8	5.5
Urban population (% of total)	13.9	34.6	5.8	12.5
Demographic characteristics				
Life expectancy (years)	37.2	49.6	37.2	44.2
Health and nutrition				
Population per physician	48,110	13,670[a]	82,170	38,790[b]
Pop. per nursing person	3,280	1,910[a]	8,460	4,650[b]
Pop. per hospital bed	540	370[c]	2,210	1,430[d]
Calorie supply per capita				
(% of requirements)	91.3	105.1	101.1	91.5
Infant mortality rate				
(per thousand)	162.5	108.8	191.2	145.7
Education				
Primary school enrollment				
ratio	65.0	104.0	5.0	23.0[b]
Secondary sch. enrollment				
ratio	2.0	18.0	0.3	5.0
Adult literacy rate	18.9	40.5[e]	0.9	9.8
Employment (%)				
Labor force participation rate	51.2	45.4	32.9	31.2
Labor force in agriculture	87.0	83.0	95.0	91.0
Labor force in industry	5.0	7.0	1.0	3.0

[a]1979.
[b]1978.
[c]1977.
[d]1976, excludes specialized hospitals.
[e]1976.

Source: World Bank, 1983: II, 16 and 68.

internally directed political and economic development. Nigeria's political and economic structures were fashioned during the colonial era (1860–1960) to serve the needs and interests of the colonial power. Its reliance, after independence, on a development strategy emphasizing export trade and capital accumulation has reinforced the structure of political and economic dependence imposed during the country's colonization. It is argued that the priority given to the petroleum sector by Nigerian leaders and planners led to greater internationalization of the economy. As the petroleum sector grew, the vulnerability of Nigerian development to world economic conditions increased. This meant that successful development of petroleum would exacerbate the society's dependence rather than foster its independence. This study concludes that the architects of Nigerian development strategy bear some responsibility for the absence of development because they failed to anticipate the internationalization and growing dependence of the society on the needs of developed economies. However, as the study demonstrates, the fundamental obstacle to Nigerian development has been a structural condition of underdevelopment that it shares with many Third World societies. Unless structural change is effected that releases Nigeria from its post-colonial dependence, development will remain elusive.

THE ROLE OF TRADE AND CAPITAL ACCUMULATION IN THE DEVELOPMENT OF NATIONS

The beginning of the industrial revolution in Great Britain changed the idea of national economic development from self-sufficient, primarily agricultural growth gained over an extended period of time to internationally based, rapid expansion of key manufacturing sectors. Hence, development theory after the British industrialization focused on trade, industry, capital, and global markets, and saw the development process in a compressed time frame.

The Developed Countries' Experiences

Since the industrial revolution in England, many nations have achieved relatively high levels of development in their economies. International trade and capital accumulation played crucial roles in the economic development of these nations. Many of the presently advanced nations underwent development processes in which the export sector was dominant or contributed significantly to economic development.

In discussing the role of primary product exports as a source of growth during crucial preconditions and takeoff stages, Cairncross (1961:236) stated:

> Whether one thinks of Britain at the outset of the industrial revolution or the United States in the nineteenth century or of Japan in twentieth, the expansion of exports gave a conspicuous momentum to the economy and helped it on its way to industrialization.

As pointed out by Matthews et al. (1982), between 1856 and 1973, exports as a percentage of GDP in Britain ranged from 14.6 percent to 25 percent. Furthermore, a comparison of the growth of the exports sector and that of the GDP shows that for most of the above period, a drop/rise in the growth rate of the exports sector was followed by a corresponding drop/rise in the growth rate of the GDP (Matthews et al. 1982:428).

TABLE 1.3. Share of World Exports of Six Developed Countries, 1881–1973
(percent, based on values in U.S.$ at current prices)

Year	USA	Germany	Italy	France	Japan	Sweden
1881–85	6.0	16.0	2.0	15.0	0.0	1.0
1899	12.1	16.6	3.8	14.9	1.6	0.9
1913	13.7	19.9	3.5	12.8	2.5	1.5
1929	21.7	15.5	3.9	11.6	4.1	1.8
1937	20.5	16.5	3.7	6.2	7.4	2.8
1950	26.6	7.0	3.6	9.6	3.4	2.8
1964	20.1	19.5	6.2	8.5	8.3	3.4
1973	15.1	22.3	6.7	9.3	13.1	3.3

Source: Matthews et al. 1982:435.

The experience of other developed nations was similar to that of Britain. Tables 1.3 and 1.4 present the share of world exports and the annual growth rates of GDP for the United States, Germany, Italy, France, Japan, and Sweden. An examination of these tables reveals that as a country's share of world exports has increased in the post-World War II period, GDP has grown as well. The best example of a developed country whose exports sector contributed significantly to its post-World War II economic development is Japan. Through its aggressive marketing of exports, it has been able to increase its share of world exports from a meager 1.6 percent in 1899 to 13.1 percent in 1973. The result has been an increase in the annual percent growth of GDP per man-year from 1.8 percent for the period 1899–1913 to 8.4 percent for 1964–73.

TABLE 1.4. Growth of GDP Per Man-Year in Six Developed Countries, 1873–1973 (annual percent growth rates)

Period	USA	Germany	Italy	France	Japan	Sweden
1873–1899	1.9	1.5	0.3	1.3	1.1	1.5
1899–1913	1.3	1.5	2.5	1.6	1.8	2.1
1913–1924	1.7	-0.9	-0.1	0.8	3.2	0.3
1924–1937	1.4	3.0	1.8	1.4	2.7	1.7
1937–1951	2.3	1.0	1.4	1.7	-1.3	2.6
1951–1964	2.5	5.1	5.6	4.3	7.6	3.3
1964–1973	1.6	4.4	5.0	4.6	8.4	2.7
1873–1951	1.7	1.3	1.3	1.4	1.4	1.7
1951–1973	2.3	4.8	5.5	4.4	7.9	3.0
1873–1973	1.8	2.0	2.4	2.0	2.6	1.9

Source: Matthews et al. 1982:31.

The Newly Developing Countries' Experiences

The developing nations of the Third World have found the trade- and capital-based path to development less successful. The question of why developing nations have not been successful in achieving economic development comparable with that of the developed countries has preoccupied those involved with theory and policy throughout this century.

Many argue that lack of capital and inadequate foreign trade have acted as constraints or bottlenecks on the economic development of Third World nations. But the 1970s proved that mere capitalization of an economy and a high volume of foreign trade are in themselves inadequate for economic development. This issue becomes clearer as we examine the experiences of the petroleum-producing developing countries (PPDCs), especially Nigeria.[1]

Following the sharp rise in the prices of petroleum during 1973–74, the PPDCs suddenly found themselves with abundant oil revenues, so that foreign exchange ceased to be a constraint on their economic development. As reported by the American Express Bank, the international current account balances (in billion of dollars) for OPEC each year from 1976 to 1982 were 36.5, 29.0, 4.5, 61.0, 109.0, 60.0, and -18.0, respectively (TIME 1983:42). As these figures show, OPEC reached a peak in the accumulation of foreign reserves in 1980 ($109 billion), but by 1982 it was faced with an estimated deficit of $18 billion. Despite substantial internally generated capital, OPEC individually and collectively has failed to exhibit a consistent pattern of development.

Because of the belief that without adequate infrastructure, a developing nation cannot bring about sustained growth, these countries with new oil wealth implemented development plans aimed at rapid growth of their infrastructure sectors. Tables 1.5 and 1.6 compare the average annual growth rates of GDP (by origin) of five OPEC countries: Iran, Iraq, Nigeria, Saudi Arabia, and Venezuela.[2] All five countries experienced tremendous growth between 1971 and 1977 in the construction, transportation and communication, and public administration and defense sectors. In addition, Iran, Iraq, Nigeria, and Venezuela reported increasing growth in manufacturing.

While the infrastructure sectors were undergoing rapid increases, growth in the agricultural sector slowed. For the selected countries, the average annual growth rate of agriculture varied from -1.5 percent for Iraq and Nigeria to 5.2 percent for Iran. This has had significant implications for the imports sector. During 1971–77, the five countries experienced an average annual growth of the imports sector ranging from 24.6 percent for Venezuela to 33.6 percent for Iraq.

The rapid growth of the imports sector of these developing nations brought substantial balance of payments problems as the surplus of foreign reserves accumulated during the mid-1970s was depleted in the latter part of the decade. Many of the PPDCs suddenly found themselves in economic chaos and facing mounting foreign debt. This dramatic turn of events suggests that, first, possession of significant capital does not by itself assure successful economic development; second, the rapid development of industrial infrastructure will result in declines in agriculture that lead to substantial trade imbalances; third, the growth of the trade, industrial, and infrastructure sectors of Third World countries is not in itself a solution to underdevelopment.

NIGERIA AND THE STRUCTURE OF UNDERDEVELOPMENT

The lack of development in Nigeria can be traced to a structural condition of underdevelopment. By "underdevelopment" is meant the persistence of stagnation, duality or lack of integration between agriculture and industry and between urban and rural spatial structures, inequality as the level of economic activity grows, and frequent political instability.

TABLE 1.5. Average Annual Growth Rates of GDP, by Origin, Iran and Iraq: 1960–77

	Iran		Iraq	
Sectors	*1960–70*	*1971–77*	*1960–70*	*1971–77*
Miscellaneous	8.5	16.7	9.5	9.5[a]
Agriculture	4.4	5.2	5.7	-1.5[a]
Mining	14.0	0.6	4.0	10.6[a]
Manufacturing	12.0	16.1	5.9	11.5[a]
Construction	8.5	12.0	5.7	23.1[a]
Electricity, gas, water	25.1	14.7	12.3	15.6[a]
Transportation, commun.	4.4	17.4	4.4	14.3[a]
Trade, finance	10.6	17.9	9.6	10.8[a]
Public admin./defense	13.3	13.6	8.8	18.0[a]
Imports	15.6	25.3	20.8	33.6[b]
Exports	21.3	36.6	36.7	53.4[b]
GDP at factor cost	11.3	7.4	6.1	10.8[a]
GDP at market cost	11.2	7.4	6.2	8.1[b]

[a]1970–76.
[b]1970–75.

Source: World Bank 1980:106–09.

Before its political independence in 1960, Nigeria's economy was colonial in nature. Since independence, Nigeria has merely moved from manifesting the structures of a colonial society to manifesting those of a neocolonial society. By a "neocolonial society" is meant one that is dependent on, and indirectly controlled and influenced by, developed economies. It is tied to the world market despite political independence. Such an economy, because of its dependent nature, produces a few key products for export, usually raw materials (agriculture goods and minerals), whose prices fluctuate widely and depend on demand from developed economies. In many cases, the resources of a neocolonial economy are those requiring heavy capital investment and advanced technology for extraction (such as petroleum), causing dependency on developed economies to exploit them.

TABLE 1.6. Average Annual Growth Rates of GDP, by Origin, Nigeria, Saudi Arabia, and Venezuela: 1960–77

Sectors	Nigeria		Saudi Arabia		Venezuela	
	1960–66	1971–77	1963–70	1971–77	1960–70	1971–77
Miscellaneous	-2.2	3.0	7.1	6.9	—	8.8
Agriculture	1.3	-1.5	1.0	3.7	5.7	3.5
Mining	31.7	8.1	11.1	13.1	3.1	-6.9
Manufacturing	12.8	13.4	10.7	4.8	6.2	6.4
Construction	9.0	24.7	8.2	23.5	3.0	14.3
Electricity, gas, water	16.7	18.2	17.2	12.9	13.1	10.7
Transport, commun.	4.0	16.5	14.5	19.0	6.0	9.4
Trade, finance	—[a]	—[a]	11.2	15.5	8.9[b]	6.7
Public admin./defense	4.8	24.6	9.0	9.6	—	8.9
Imports	4.0	24.7	27.3	26.8	18.9	24.6
Exports	12.9	5.6	58.7	76.4	29.4	34.0
GDP at factor cost	4.3	6.2	9.9	12.7	5.9	5.6
GDP at market cost	4.4	6.0	9.9	12.7	5.9	5.6

[a]Included in Miscellaneous.
[b]1960–67.

Source: World Bank, 1980:150–51, 210–11, 226–27.

Nigeria failed to develop because it did not undergo, and has not undergone, structural change. The social, political, and economic character of the society has not been transformed, but continues to resemble its colonial form. The political organization of Nigeria remains bureaucratic and elitist in the mold of its colonial administration. Its economy is dependent upon a single trade sector. While the trade commodity has changed from agricultural products to petroleum, the essential structure of the economy is unchanged.

It is argued in this study that in order to break the condition of underdevelopment, Nigeria must undergo structural change that will create conditions for autonomous development. Autonomy is a means to other social aims that should guide the development process, such as greater equality and a just and egalitarian society. Without autonomy, a country's ability to progress along dimensions such as justice is greatly limited.

By "autonomous development" is meant development that is internally controlled, directed, and influenced. It involves self-determination without outside dominance. Such a mode of development has as its goal the achievement of economic and political self-reliance. But self-reliance does not mean self-sufficiency, as pointed out by Russell Anderson:

> Self-reliance does not mean isolation, nor is it equivalent to self-sufficiency. Self-reliance is development which stimulates the ability to satisfy basic needs locally: the capacity for self-sufficiency, but not self-sufficiency itself. Self-reliance represents a new balance, not a new absolute. (As quoted in Morris 1982:137)

In the context of Nigerian development, this means that Nigeria strives to achieve a new balance between its internally directed production system and social needs such that the provision of basic goods and services cannot be threatened by external economic and political forces. It does not mean that Nigeria must be self-sufficient in all production, nor does it mean that Nigeria must close its door to all outside influence. Instead, it means that Nigeria must establish a degree of self-sufficiency in certain areas of production and must be able to decide for itself the nature and pace of technology transfer, world trade involvement, and foreign capital attraction.

Without a capacity for autonomous development and self-determination, a country such as Nigeria will be a passive participant in determining its own future. If Nigeria is to achieve autonomy and self-determination, economic growth will not be enough, however substantial that growth is. Structural change must occur if growth is to be translated into development. In other words, Nigeria must have both growth and change.

METHODOLOGY

The study of national development, as Myrdal has observed, requires an understanding of a country's history, politics, theories, ideologies, economic structures and levels, social stratification, agriculture, and industry (Myrdal 1968). As he cautioned, the study of these dimensions of development cannot take place in isolation, but must be viewed "in their mutual relationship." Hence, this study of Nigeria provides a multidimensional analysis of development. The Nigerian development experience is considered as the interrelation of the economic development paradigm that guided development thinking, the national planning model employed, the ideology of nationalism, and the internal and external factors in Nigeria's environment that combined to influence its development strategies.

Figure 1.1 depicts the environment in which Nigerian development has taken place and continues to take place. According to this figure, the ideology of nationalism, the national planning apparatus, and the economic development paradigm followed by Nigeria's leaders and planners all have direct effects on its development strategies. Furthermore, the ideology of nationalism affects the choice of planning orientation and structure. The choice of development paradigm affects the choice of planning models.

The ideology of nationalism, the competing theories of development, the development planning orientation, and certain environmental factors together provide critical insight into the strategic development choices of Nigeria and express the basis of her hopes to date for socioeconomic development. Through them, we can begin to understand how Nigeria has come to depend upon the petroleum economy and its revenues, why it permitted its agricultural sector to decline, why it pursued a policy of rapid development of industrial and physical infrastructure, and how it hoped to moderate foreign economic influence through what it called indigenization of industry. We can also understand how and why Nigeria came to inherit its current social, economic, and political problem.

Three methodological procedures are used in this study to analyze the factors that shaped Nigerian society and guided its development choices.

First, an attempt is made to locate Nigerian development strategy within the theories and ideas of development. The effects of neoclassical ideas, development planning models, and the ideology of nationalism on Nigerian thinking are examined.

Second, a sociopolitical and socioeconomic analysis of Nigeria is developed that assesses the nature and extent of social change in Nigeria from independence to the present. This analysis is developed in three stages: the colonial period, a transitional period from 1960 to 1970, and the oil-boom era to the present.

Finally, a structural analysis of Nigeria's development experience is conducted in an effort to evaluate the extent to which that experience can be explained by internal political, economic, and social factors, and the extent to which it reflects and is the result of a condition of underdevelopment. The role of Nigerian nationalism, lack of political development, the rise of the military, the Biafran civil war and its aftermath, and widespread corruption are considered. Then the role of Nigeria's colonial past and its export petroleum economy in blocking her development is explored.

FIGURE 1.1   Relationships between Ideology of Nationalism,
Economic Development Paradigm, Development Planning Models, Internal and External Factors, and Development Strategies in Nigeria

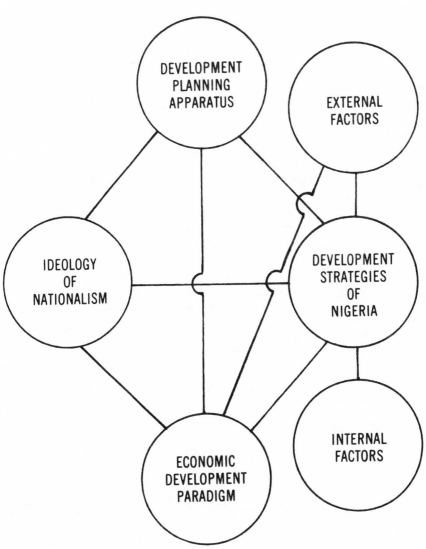

Source: Provided by author.

OUTLINE OF CHAPTERS

Certain ideas of development greatly influenced development strategy in Nigeria. These ideas were drawn mainly from neoclassical economic development theory. Chapter 2 identifies the development paradigm followed by Nigeria, which is rooted in the classical and the neoclassical economic theories. It discusses how this paradigm influenced Nigeria's understanding of the development process and what could be done to effect it.

Because neoclassical theory was written to explain the experience of developed nations, there arose a body of criticism that challenged neoclassical ideas. Some of these criticisms were recognized and accommodated by Nigerian development planners, but many were not. Chapter 3 presents the ideas of Third World development theorists and their structural criticisms of neoclassicism. These ideas are used in Chapter 8 to analyze Nigeria's development strategy.

Chapter 4 analyzes the role of the ideology of nationalism, and examines the development planning orientation chosen by Nigeria, particularly the choice of a development planning apparatus and its evolution in Nigeria.

Chapter 5 addresses Nigeria's drive to independence. It contains an analysis of the country's social, political, and economic development from the period of colonialism to 1960.

The 1960s were years of extraordinary changes for Nigeria. Chapter 6 analyzes that period of transition, focusing on Nigeria's development planning and strategies, the types of economic and sociopolitical changes that took place, and the rise of the petroleum economy. As the chapter brings out, the First National Development Plan (1962–68) was drawn by technocrats trained in neoclassicism.

In retrospect, the 1970s may be remembered as the era during which Nigeria missed its opportunity to take off into sustained growth. Chapter 7 covers that decade of oil boom in Nigeria, and beyond it to 1985. The focus of the chapter is on the growth resulting from the oil boom, the impacts of petroleum revenues on the other sectors of the Nigerian economy, and the implications of the petroleum-export orientation. Again, the reliance of the national planning apparatus on neoclassical development ideas is shown.

In Chapter 8, a critical analysis of Nigeria's costly progress is presented. This chapter analyzes and interprets the findings on development in Nigeria from 1960 to 1984. The central issue is whether Nigeria could have anticipated its crisis and taken steps within its economic and political structures to have avoided it, or whether those structures are themselves the root problem.

Chapter 9 contains the conclusions derived from the research. It makes recommendations regarding future courses of action on the management of internal resources and the pursuit of national development in light of Nigeria's condition of underdevelopment. The chapter also discusses the lesson provided by the Nigerian experience for understanding the nature and limits of development in the Third World, especially development predicated on capital accumulation and export trade. Finally, it presents an agenda for a new Nigeria.

NOTES

1. A petroleum economy can be defined by three characteristics. First, a major proportion of the government's expenditure is financed from oil revenues (see Table 1.7).

TABLE 1.7. Foreign Trade Structure: Agricultural and Mineral Products as Percentage of Total Exports, Selected OPEC Countries, 1960–80

Country	Food and Nonfood Agric. Exports				Petroleum and Other Minerals			
	1960	1965	1970	1980	1960	1965	1970	1980
Iran	10.0	8.1	6.2	1.8[a]	87.5	87.5	89.8	96.5[a]
Iraq	7.2	4.0	4.2	0.8[a]	92.3	94.9	94.9	99.2
Nigeria	81.1	65.3	36.4	4.1[b]	11.4	32.4	62.3	95.3[b]
Saudi Arabia	0.0	1.0	0.1	0.1	99.6[c]	97.7	99.8	99.2[b]
Venezuela	1.4	1.0	1.7	0.4	92.5	97.2	96.8	97.9

[a]1976.
[b]1979.
[c]1962.

Source: World Bank, 1980:396–401; 1983:518–22.

Second, the petroleum sector is a main contributor to the GDP even though it employs a very small proportion of the nation's labor force. Third, the nonpetroleum sectors of the economy are price-takers in the market for imported goods and services (Alam 1982).

2. Significant variation can be found among the economies of PPDCs. Hence, it is difficult to compare them and generalize Nigeria's experience to all of them. Petroleum economies in PPDCs can be grouped into two categories. First, here are the economies of Kuwait, Saudi Arabia, the United Arab Emirates, Oman, and Libya, where there is a high inflow of petroleum revenues relative to population. This is the Type I petroleum economy. The second category is the economies of Algeria, Indonesia, Iran, Mexico, Nigeria, and Venezuela, where the per capita inflow of petroleum revenues is relatively low. This is the Type II petroleum economy. It is reasonable to expect that in a period of falling petroleum prices and the accompanying sharp drop in oil revenues, the Type II economy is more likely to face severe crisis.

2 The Neoclassical Development Paradigm and Nigerian Development Strategy

INTRODUCTION

In order to understand the Nigerian case, we need to understand, first, the perspectives from which Nigeria had the opportunity to choose and which shaped its development strategy. It is argued that a set of perspectives, drawn primarily from neoclassical economic theory, influenced both policy making and the formulation and execution of the various National Development Plans in Nigeria. (The plans are analyzed in Chapters 5–8.) This book examines the literature for a particular purpose: to lay out the development paradigm that Nigeria adopted.

Development strategies in Nigeria relied heavily on classical and neoclassical ideas about the basis for national development. These ideas stress the roles of trade and capital accumulation in development. While they differ in the way they analyze these factors, they are in basic agreement on the centrality of these factors in determining development.

CLASSICAL INSIGHTS

Economic development theory, like every other branch of social theory, is greatly affected by the time in which it is conceived. The economists of the late eighteenth and nearly nineteenth centuries were very much interested in the conditions conducive to industrial economic progress. They lived through the period of the industrial revolution in Europe, and thus witnessed the takeoff of sustained urban-centered industrial growth there. Their observations should be of considerable interest to us, because what happened in 18th- and 19th-century Europe is what is often proposed as a model for the Third World.

Adam Smith, the founder of the classical school, was the first to outline the general nature of economic development. In *The Wealth of Nations* he advanced the general proposition that the growth of national wealth depends, first, on the productivity of labor associated with the technologically determined "division of labour" and, second, on the accumulation of capital associated with the institutionally determined extent of "parsimony."

He referred to trade as an exchange of surpluses of commodities in excess of domestic consumption. He emphasized that foreign trade, by widening the scope of the market, improves the division of labor and thus raises productivity within the exporting country:

> Between whatever places foreign trade is carried on, they all of them derive two distinct benefits from it. It carries out that surplus part of the produce of their land and labour for which there is no demand among them, and brings back in return for it something else for which there is a demand. It gives value to their superfluities, by exchanging them for something else, which may satisfy a part of their wants, and increase their enjoyments. By means of it, the narrowness of the home market does not hinder the division of labour in any particular branch of an art or manufacture from being carried to the highest perfection. By opening a more extensive market for whatever part of the produce of their labour may exceed the home consumption, it encourages them to improve its productive powers, and to augment its annual produce to the utmost, and thereby to increase the real revenue and wealth of the society. These great and important services foreign trade is continually occupied in performing, to all the different countries between which it is carried on. (Smith [1776] 1937:415)

We can summarize Smith's discussion into two theories. First, foreign trade, by expanding the extent of the market, improves the division of labor and increases productivity within the exporting country. Second, foreign trade provides an outlet for surpluses of certain products above their internal consumption. These two theories are what Hla Myint (1958; 1977) called the "productivity" theory and the "vent for surplus" theory, respectively.

Malthus is so closely associated with the concept of population pressure that his two-sector analysis of underdevelopment is often overlooked. The "population explosion" in the last two decades of the 18th century stimulated great interest in the Malthusian population theory. And today, population pressure in the developing countries has revived its popularity. In proposing the two-sector analysis of underdevelopment, Malthus ([1836] 1951) conceived the economy as consisting of two major sectors: industrial and agricultural. In his view of economic development, capital is invested in agriculture until all the cultivable land is cultivated, stocked, and improved. After that, the opportunities for profitable investment in that sector cease to exist. At this point, opportunities for investment exist only in the industrial sector. Therefore, to avoid diminishing returns to increased employment on the land, the following conditions must exist. First, technological progress in the industrial sector must be rapid enough. Second, enough investment must take place to absorb most of the population growth in the industrial sector and to reduce the cost of living of the workers on the land, permitting reduction in their wage rates (Malthus [1836] 1951).

Malthus made some suggestions about sectoral interaction in underdeveloped areas, explaining why they remain underdeveloped. He pointed out that each sector constitutes a market for the output of the other sector (in the absence of international trade). Thus the failure of either sector to expand tends to impede the growth of the other. Growth in both sectors is essential in order to have growth at all (Malthus [1836] 1951).

Stressing the importance of international trade, Malthus said:

> Without sufficient foreign commerce to give value to the raw produce of the land; and before the general introduction of manufacturers had opened channels for domestic industry, the demands of the great proprietors for labor would be very soon supplied; and beyond this, the laboring classes would have nothing to give them for the use of their land. (Malthus [1836] 1951:342)

Marx contributed to the theory of economic development in three respects: first, in the broad respect of providing an economic interpretation of history; second, in the narrower respect of specifying the motivating forces of capitalist development; and third, in the respect of suggesting an alternative path of planned economic development (Kurihara 1959).

In his *Critique of Political Economy,* Marx outlined his conception of historical evolution, according to which economic institutions, while they are products of social evolution, are themselves capable of influencing the course of social progress. In his major work, *Capital,* Marx sees the process of capitalist development as a function of profit takers' savings at the expense of wage earners' consumption; of the investment of those savings in new capital, propelled by the competition for profits but inhibited by the disappearance of investment opportunities as a result of both underconsumption and overinvestment; and of the sociologically given size and technologically given productivity of the labor population. However, it is Marx's perception of planned development expressed in his other writings that has had a major impact on the actual economic development of the Soviet Union and China. His notion of state ownership of the instruments of production as a historically necessary condition of planned development appeals to developing nations that are trying to industrialize at a fast pace (Roll 1956).

The ideas of the classical theorists that influenced Nigeria's development strategy were international trade, capital accumulation, and state ownership of instruments of production. There may be a contradiction in accepting the market-oriented development mode and the state ownership and control of productive resources; Nigeria like other developing nations, has chosen a mixture of both (see Chapter 4).

REVISED CLASSICISM

In the aftermath of Marx's criticism, classical doctrine found itself at a crossroad. Theorists like Joseph Schumpeter sought to revise classicism while retaining the insights of Smith, Ricardo, and Marx. This approach will be discussed here. The other approach, which abandoned Marx's insights and built on the theories of Smith, will be discussed later.

Schumpeter, in his *The Theory of Economic Development* (published in German in 1911 and in English in 1934), was the first to separate economic development as a specialized area for study and analysis. His ideas can be found in the economic nationalism that swept Africa, Nigeria in particular. As discussed in Chapters 6–8, Nigeria's planning efforts were focused on trade and capital accumulation in order to stimulate rapid capitalist development; yet its leaders were convinced that capitalism was defective with respect to distribution and equity questions. This led them to temper capitalism with both state and Nigerian capitalist ownership of productive resources.

Like earlier theorists, Schumpeter's interpretation of the economic development process was strongly influenced by the economic conditions of his time. According to Schumpeter (1942), the path to capitalist development is jerky rather than steady, owing to the erratic interaction of the entrepreneur (innovator), innovation, and credit, and is therefore destined to give way to some form of socialism. This is because of its successful achievements, such as mass production, mass education, and big business, as well as the inevitable accompaniments of government regulations, intellectual resentment, and trade unionism. He emphasized the role of the entrepreneur as an innovator who is willing to take the risk of trying out modern inventions for technical improvements in production. The profit he makes is his reward for his initiative and for being the first to demonstrate the benefits of the new technology. This profit attracts new capital to production. However, the profit vanishes as soon as competitors start to imitate the innovations. Therefore, without other innovations or technological progress, excessive capital accumulation and competition result in diminishing returns. According to Schumpeter:

> Capitalism, being essentially an evolutionary process, would become atrophic. There would be nothing left for entrepreneurs to do. They would find themselves in much the same situation as generals would in a society perfectly sure of permanent peace. Profits and along with profits the rate of interest would converge toward zero. (Schumpeter 1942:139)

Nonetheless, he saw capitalism as the most rapid mode for economic development. This is because capitalism gives full reign to the entrepreneur in the development cycle while convinced of its power to encourage development. Schumpeter shared Marx's conviction about the "inevitable" breakdown of capitalism. With rapid development successfully concluded, capitalism eliminated, in Schumpeter's view, any further need for itself.

NEOCLASSICAL INSIGHTS

Neoclassical thinking, which emerged in the last quarter of the 19th century and the beginning of the 20th, reflects a defensive ideology, a need for circumventing the difficulties that had been created by the theory of labor value. The theory of labor value had been used as a weapon against capitalism. It was therefore necessary to justify the existing social order as the one permitting the most rational use of available resources by creating a new analytic instrument.

The architect of the neoclassical synthesis was Alfred Marshall. Central to his thought is the theme that economic evolution is gradual and continuous on each of its numberless routes:

> Economic conditions are mostly the result of slow and gradual development; and, partly for that reason, they commonly show the One in the Many, the Many in the One. It is necessary to look backwards a little, in order the better to look forwards.
>
> Economics is concerned mainly with general conditions and tendencies: and these as a rule change but slowly, and by small steps . . . thus the maxim that "Nature does not willingly make a jump" (Natura abhorret saltum) is especially applicable to economic development. (Marshall 1919:5–6)

The other theme that is applicable to our analysis of Third World development in general and Nigeria in particular is the One in the Many, the

Many in the One. It is Marshall's belief that, partly as a consequence of the gradual and slow nature of economic development, economic conditions and tendencies that prevail at any time and place reflect the habits of action, thought, feeling, and aspiration of the whole population. And because each affects the character of the population, the One is seen in the Many. Conversely, each tendency comprises, in some degree, almost every influence that is prominent then and there. And inasmuch as a full study would incidentally represent a complete picture of the whole, the Many are seen in the One. The important point to bear in mind, therefore, is that even though the present never reproduces the past, the past lives on after we have forgotten about it; and that the most advanced people retain much of the substances of earlier habits of associated action in industry and trade, even when the forms and names of those habits have been changed under new conditions. Marshall therefore suggested that a chief purpose of every study of human action should be to suggest the probable outcome of present tendencies, and to indicate clearly the modifications of those tendencies that might further the wellbeing of mankind (Marshall 1919).

On international trade, Marshall argued that the direct gain a country derives from external trade is the excess of real value of the products and services imported over the value of the things that could have been produced domestically by the labor and capital needed for producing the exports and the costs of administering the trade. In other words, the gain from international trade is the superior value of what is received over what is given up. He believed that genuine trade commonly benefits both parties to it, because, though each receives only what the other gives up, what is received is more desired than what is given up (Marshall 1919).

It is also worthwhile to note Marshall's interpretation of Darwin's "law of survival of the fittest" (Marshall 1919:175–77). This law, according to Marshall, means that "those races are most likely to survive, who are best fitted to thrive in their environment: that is, to turn to their own account those opportunities which the world offers to them." He stressed the role of social qualities in overcoming the difficulties that lie in the way of development—even to overcome human enemies; that those institutions tend to survive which have the greatest faculty for utilizing the environment in developing their own strength. Also, inasmuch as they in return benefit the environment, they strengthen the foundations of their own development and increase their chance of surviving and prospering. With regard to development theory, the central achievement of Marshall was to reestablish the emphasis of the classical theorists on trade and capital accumulation as fuels for growth in a country such as Nigeria.

RESURGENT NEOCLASSICAL-INTERNATIONAL
TRADE THEORY

International trade theorists such as Gottfried Haberler, Jacob Viner, Alex Cairncross, and Peter Bauer expounded and reinforced the classical and neoclassical views. This group shares the neoclassical belief that trade can promote the growth of the rest of the economy, and that an expansion of trade in any type of product produces spillover effects that are favorable to the other sectors. For example, Cairncross argues:

> I confess to some skepticism about the supposed ineffectiveness of foreign trade in producing innovation and development. It does not strike me as entirely plausible to speak as if foreign trade could be contained within an enclave without transmitting its dynamic influences to the rest of the economy. (Cairncross 1961:240)

Cairncross also emphasized the favorable "educative effects" of trade that spread aspirations, skills, and technical knowledge.

Haberler emphasized that international trade has several indirect and dynamic benefits, apart from static gains to the trading countries. In his view, international trade has made a tremendous contribution to the development of less developed countries (as well as of most developed ones) in the 19th and 20th centuries, and can be expected to make an equally substantial contribution in the future, if it is allowed to proceed freely (Haberler 1968). According to him, international trade has several benefits:

> First, trade provides material means (capital goods, machinery and raw and semi-finished materials) indispensable for economic development. Secondly, and even more important, trade is the means and vehicle for the dissemination of know-how, skills, managerial talent and entrepreneurship. Thirdly, trade is also the vehicle for the international movement of capital especially from the developed to underdeveloped countries. Fourthly, free international trade is the best antimonopoly policy and the best guarantee for the maintenance of a healthy degree of free competition. (Haberler 1968:108–09)

Using petroleum as an example of a primary export commodity that has enjoyed a tremendous level of world demand, Haberler states:

> it is the ancient rule again: Those who have (oil deposits) shall receive (foreign capital). If a wide range of primary commodities other than

crude oil were enjoying an equally strong increase in world demand, is there much reason to doubt that a larger volume of private capital would be attracted to the underdeveloped countries, in spite of the political risks which, in varying forms and degrees, have always existed and will always continue to exist? (Haberler 1968:101)

Peter Bauer (1959:112) had the following observation: "But is it not the case that now, as in the past, the most advanced of the underdeveloped regions and sectors are those in contact with developed countries?" As we can see, the argument of this group builds upon the classical and neoclassical views and, therefore, their conclusions are similar.

The ideas of classical and neoclassical economists are, in a nutshell, contained in a quotation from Dennis Robertson (1947:501):

The specializations of the nineteenth century were not simply a device for using to the greatest effect the labours of a given number of human beings; they were above all an engine of growth.

To further support the belief of these economists that production for international markets could successfully foster economic development, Hla Myint (1958:318–19) stated:

The "productivity doctrine" looks upon international trade as a dynamic force which, by widening the extent of the market and the scope of division of labor, raises the skill and dexterity of the workmen, encourages technical innovations, overcomes technical indivisibilities and generally enables the trading country to enjoy increasing returns and economic development.

Gerald Meier (1968) mentioned "the education effect" of trade expansion on a country's economic development. This includes several influences, ranging from incentives to produce resulting from the desired consumption of imported goods, to the workings of the international transmission of skills, knowledge, know-how, and technology. Other classical and neoclassical economists argue that international trade based on comparative advantage results in an efficient allocation of resources (Kramer et al. 1966:47–70). The view of traditional theorists was nicely summed up by Meier (1968:222):

If trade increases the capacity for development, then the larger the volume of trade the greater should be the potential for development.

In other words, the larger the growth in international trade, the larger the growth of the country's economy, irrespective of the type of export products or the exporting country's pattern of trade and stage of development.

As Chapters 5–7 show, Nigerian policy makers adhered closely to the above view in devising their development plans.

STAGE THEORIES

Marshall's development theory is essentially microeconomic in character, focusing on the interplay of economic forces in individual markets and stressing the decisions of individuals and firms in yielding an orderly and incremental pattern of growth. Stage theories represent an effort to give neoclassical insights on development a macroeconomic orientation.

W. W. Rostow in his *The Process of Economic Growth* (1962) summarized economic growth possibilities in six propensities that reflect a society's response to the challenge of the economic environment. These are the propensity to develop fundamental science; the propensity to apply science to economic ends; the propensity to accept innovations; the propensity to seek material advance; the propensity to consume; and the propensity to have children. Directly and indirectly, these six propensities are believed to be the determinants of the rate of growth of the quantity and quality of the labor force and of the capital stock of a society. According to Rostow, the path of economic growth can be divided into four functional stages. First is the traditional society, where the preconditions for the second stage are set. The second stage is the takeoff. The third stage is the drive to maturity, during which sustained growth consolidates itself. The final stage is the period of high mass consumption. The essential link in the process, according to Rostow, is the takeoff stage, for which he lists three common characteristics: a rise in the rate of productive investment from, say, 10 percent or less to over 10 percent of national income (or net national product); the development of one or more substantial manufacturing sectors with a high rate of growth; and the existence or quick emergence of a political, social, and institutional framework that exploits the impulses for expansion in the modern sector and the potential external economy effects of the takeoff, and gives to growth an ongoing character (Rostow 1963:164).

Following the takeoff is a long interval of sustained if fluctuating progress with 10 to 20 percent of the national income steadily invested, permitting output regularly to outstrip the increase in population. This period is what Rostow called the "drive to maturity," maturity being generally attained some 60 years after takeoff. The interval between takeoff and maturity varies from one country to another. For Great Britain, Rostow assigned the years 1783–1802; for France, 1830–1860; for Belgium, 1833–1860; for the United States, 1843– 1860; for Germany, 1850–1873; for

Sweden, 1878–1900; for Japan, 1878–1900; for Russia, 1890–1914; for Canada, 1896–1914; for Argentina, from 1935; and for Turkey, from 1937. Rostow summarized his stage theory in the following manner:

> The take-off is defined as an industrial revolution, tied directly to radical changes in methods of production, having their decisive consequence over a relatively short period of time.
>
> This view would not deny the role of longer, slower changes in the whole process of economic growth. On the contrary, take-off requires a massive set of preconditions going to the heart of a society's economic organization and its effective scale of values. Moreover, for the take-off to be successful, it must lead on progressively to sustained growth; and this implies further deep and often slow-moving changes in the economy and the society as a whole.
>
> What this argument does assert is that the rapid growth of one or more new manufacturing sectors is a powerful and essential engine of economic transformation. Growth in such sectors, places incomes in the hands of men who will not merely save a proportion of an expanding income but who will plough it into highly productive investment; it sets up a chain of effective demand for other manufactured products; it sets up a requirement for enlarged urban areas, whose capital costs may be high but whose population and market organization help to make industrialization an on-going process; and finally, it opens up a range of external economy effects which, in the end, help to produce new leading sectors when the initial impulse of the take-off's leading sectors begins to wane. (Rostow 1963:185–86)

If the assumptions of "takeoffs" and "big pushes" are valid, then an underdeveloped country will achieve development only if it undertakes the effort on a fairly massive scale. Such an offensive may require simultaneous changes in various areas of the economy, or it may require a very heavy concentration of efforts on crucial leading sectors. In either case, a major assault on its problems is required by a developing country; otherwise, the "vicious circle of poverty" will continue.

The question to ask, then, is who will be able to engineer such an effort? How can such an offensive be launched in developing nations? Even Rostow's colleagues have attacked the stage theories on this point. For instance, Prof. Henry Rosovsky called these theories "mish-mash" (Rosovsky 1965).

Simon Kuznets (1965) in his article, "Notes on the Take-off," emphasizes the fuzziness in the way the takeoff stage was delimited and in the way the distinctive characteristics were formulated. He points out that much of what Rostow attributes to the takeoff stage had already occurred in the precondition stage. He also found the stage of self-sustaining growth following the takeoff "somewhat of a puzzle. Is it self-sustained in a sense in which it is not during the take-off and/or any earlier phase?" (1965:230). He pointed out that the term "self-sustained growth" should be avoided because it is an analogy rather than a clearly specified characteristic (Kuznets 1965:231). According to him, statistical evidence for those countries on Rostow's list has failed to confirm the doubling of capital investment and the implicit sharp acceleration in the rate of growth of national product, claimed by Rostow as major characteristics of his takeoff period. Also, there is no clear distinction between the precondition and the takeoff stages; and the analysis of those stages failed to consider the effect of historical heritage, time of entry into the process of modern economic growth, degree of backwardness, and other characteristics of the early phases of modern economic growth in the various traditional countries. Finally, the concept (and stage) of self-sustained growth is seen as a misleading oversimplification because no growth pattern is purely self-sustaining or purely self-limiting.

However, as Prof. Benjamin Higgins (1968:186) rightly puts it:

One thing, however, is clear; no matter how critical Rostow's colleagues may be of his system, his terminology is here to stay. The expressions, "the take-off" and "self-sustained growth," are thoroughly entrenched in the literature, and will continue to be used by development economists, including the present writer.

Perhaps the impact of these theories can be seen most clearly in the United Nations and its specialized agencies, particularly the World Bank, which seeks to promote a staged development pattern: first, increase agricultural production; then develop village industries or consumer goods industries (import-substitution industrialization); follow this with social overhead capital projects undertaken by the government; and finally, industrialization proper. Nigeria, as is shown in Chapters 5–8, pursued the strategy that was consistent with the ideas of staged growth. It sought first to stimulate export agriculture, then petroleum export. Import-substitution industrialization was undertaken, followed by heavy investments in social overhead capital projects. In this respect, Nigeria's strategy, while taking advantage of stage theories, was consistent with the neoclassical insights that emphasize capital accumulation and trade as engines for development.

BALANCED AND UNBALANCED GROWTH

A strategic question from the perspective of the neoclassical paradigm is the type of balance to be maintained between the different sectors of the economy. Some questions that arise are: Should a country such as Nigeria give special attention to agriculture? to industry? Or should the country try to achieve a balanced growth in which all the major sectors of the economy develop side by side? How should priorities be determined?

There are strong arguments on each side. Those who favor unbalanced growth (that is, giving special priority to a particular sector) might say that a developing country has only limited resources to give to investment and growth. Therefore, it cannot do everything at once. It must choose those areas which promise the greatest development. Such areas typically do not include agriculture; economic development and industrialization are supposedly synonymous phenomena.

Those opposed to the unbalanced growth approach, and particularly those who favor a more balanced approach in which agriculture and industry develop in a complementary relationship, respond that efforts to develop industry and not agriculture have consistently failed in the Third World. In part, this is because industry depends on agriculture for the raw materials and other inputs that make production possible. Moreover, as the economy develops, the accompanying population increase will raise the demand for food and other essential commodities. As is shown in Chapter 7, a country such as Nigeria, because of lack of necessary inputs, may experience idle industrial capacity, which may ultimately result in reduced employment and scarcity of essential commodities. The concomitant need to import food and other commodities will tend to cause strain in the country's international balance of payments.

Apart from the above arguments, economists have recently concluded that, contrary to static equilibrium theory, the process of economic development is far from being as continuous, gradual, and incremental as classical theory and much general thinking on development phenomena appear to assume. Nor do they seem as ready to expect that the road to development followed by England, Germany, the United States, and other developed countries can be prescribed for the developing countries of today. The route toward development followed by different countries does not have a uniform internal structure dictated by universal principles; instead, it is a sequence of events, each partly determined by what has occurred in the world economy, and partly by internal conditions of social, political, and economic organization. In addition, the existing international economic structure influences the prospects and direction of a country's development (Schiavo-Campo and Singer 1970).

From the realization that economic development means growth and change, and the fact of the magnitude of the tasks required, as well as the complexity and inertia of existing economic and social structures, it has become evident to at least some in the international economic community that the gradual and almost unplanned approach is not realistic. Distinct planning perspectives have emerged around the concepts of "balanced growth" and "unbalanced growth." The basic contention of both is the same: that the development process is a series of discontinuous jumps. In this respect, both are innovations on the stage theory that attempt to incorporate the insights of the social development economists.

Balanced Growth

This approach views discontinuities in the economic development process as the basis for a large-scale frontal attack to keep the economy moving. The principal contributors to this view are W. A. Lewis, Ragnar Nurkse, Paul N. Rosenstein-Rodan, and Tibor Scitovsky. In brief, this economic thought stresses either the need for avoiding imbalances in supply or the need for ensuring diversified demand. Because of important differences between demand-centered and supply-centered balanced growth approaches, it is useful to discuss the two versions separately.

The supply version views economic development as requiring a balanced growth of the individual production sectors of the economy; if the output of a production process is to rise, some increase in the inputs is required. The demand for inputs will grow until all intermediary industries are producing at full capacity. For this process to proceed efficiently, a balanced approach to growth is necessary. In this sense, this version seeks a supply or production balance among sectors. As might be expected, economists supporting the supply version find utility in national planning. It stresses the need for parallel expansion of direct inputs and outputs and the need for avoiding sectoral imbalances. While this may be termed a prescription for economic growth, it is inadequate as an economic development strategy because it would all but eliminate the likelihood of large-scale structural change accompanying development.

The demand version springs from the concept of external economies. Here, the issue is how to ensure that the benefits of a large-scale national production process accrue to the wider society. Even though particular investments may be profitable to the nation as a whole, they may not be profitable to an individual investor, and therefore may not be undertaken without government intervention in the investment process. Examples are new roads and railroad tracks. Such investments add to the profitability of all producers using them. Rosenstein-Rodan (1957; 1963) expounded this view in his three-way classification of economic indivisibilities. First is the indivisibility in the production function or lumpiness of capital in the supply of social overhead capital (power, transport, communications, housing, energy). These require substantial initial investment. It is argued that in the early stages of development, excess capacity may be unavoidable because these utilities cannot be continuously scaled. Furthermore, investment in social overhead capital often must take place before other productive investment, making the appropriate scale problematic.

The second indivisibility is that of demand. An investment frequently will be too risky if there are no complementary projects. For many capital projects, it is essential that substantial external economies occur to producers from the existence of internal scale economies in another sector that manufactures intermediate products to be used by these producers.

The third economic indivisibility is in the supply of savings; it points out that there are kinks in the savings supply curve. As with social overhead capital, a critical minimum level of savings is required to give industries the threshold they require to take off successfully. Also, there is need for a synchronized application of capital to a variety of industries. This is the only way to overcome the problem of the small and inelastic demand in a low-income country. It is quite a shift from the traditional tendency to regard a developing country as a market or as a raw materials source for advanced economies. By investing in several projects simultaneously, the complementarity of demand will reduce the risk of not finding a market. In fact, new producers would be each other's customers, thereby creating a diversified demand.

As is shown in Chapters 5–7, Nigeria's development planning did not follow the approach of a balanced growth that advocates a "big spread," a strategy by which development should branch simultaneously in several directions. Rather, Nigeria chose the unbalanced growth approach that is discussed below.

Unbalanced Growth

The approach called "unbalanced growth" proceeds from the premise that development is fostered by optimally unbalanced sequences of sectoral investment. The principal exponents of this approach are Hans Singer (1949; 1958) and Albert Hirschman (1958). They and other opponents of the "balanced growth" approach have criticized it on several grounds. First, as Singer pointed out, in underdeveloped countries the initial resources necessary to simultaneously invest in several sectors (the frontal attack) are generally lacking, especially sufficient amounts of domestic capital. Second, this approach may result in a dualistic economy, superimposing a modern public sector upon a private subsistence sector.

Hirschman contended that a policy of deliberate unbalancing of the economy in accordance with some predetermined strategy is best suited for economic development. He identified the "ability to invest" as the more relevant and more serious bottleneck (Hirschman 1958:36). Thus, "ability to invest" is required in order for the type of simultaneous investment advocated by the "balanced growth" group to take place. Ability to invest, however, depends on the size of the domestic market—which in underdeveloped areas is small. In sum, balanced growth will require large quantities of precisely the scarce resource lacking in a developing country: investment capital. The correct approach, according to Hirschman, is to economize on the scarce resource, ability to invest (investment capital), by maximizing induced investment decisions and to restrict autonomous investment decisions to those strategic sectors where they are most necessary to initiate a process of growth. Induced investment is capacity-oriented investment, occurring in response to changes (increases) in consumer demand and changes in output. Autonomous investment, on the other hand, is independent of current consumer needs and is made to tap a potential desire for a (new) product or service; it is usually associated with innovation.

Hirschman feels that growth is more likely to be faster with chronic imbalances set up in the system because of the incentives and pressures they provide. For example, it is possible to concentrate investment on social overhead capital such as roads, communications, electricity, and hospitals, and believe that investments in directly productive activities will follow as a result of the increased profitability of this sequence (a "permissive" sequence). Also, a "compulsive" sequence may occur where social overhead capital is neglected in favor of directly productive investment in the hope that this will create a pressure to build overhead capital.

From the above, it becomes clear that by pursuing a policy of unbalanced growth, a country has put itself in a situation whereby further progress is either permitted or compelled by the interdependence of events. Hirschman's preference is for a compulsive sequence. But, depending on the structure of the country, different sequences may be in order, and therefore he leaves the determination of the specific strategy of development to the individual case. His main principle is to invest in those projects that have the greatest total linkages, forward and backward, as measured by input-output coefficients, and to gear national and regional planning to supporting those projects. One could criticize Nigeria for choosing the unbalanced growth strategy. Nigeria, however, did not have timely and reliable statistical data to implement either of the approaches. Therefore, for Nigeria, of the two, unbalanced growth is the simpler.

SUMMARY

In sum, the development paradigm followed by Nigeria was rooted in the neoclassical theory of economic growth, with modifications that recognized indigenous characteristics and problems. Its strategy was built around capital accumulation and trade. Nigeria sought to follow a capitalistic development path, but its colonial past and its multiethnic social structure (both discussed in Chapter 5) led it to seek and adopt a policy of restraints on unbridled capitalism. In this respect, Nigeria recognized the tensions in capitalist development identified by Schumpeter.

In emphasizing trade as a fuel for development, Nigeria accepted the Marshallian view that involvement in world markets would be to its advantage because, purportedly, international trade benefits both parties to it. Nigeria sought to institutionalize growth through trade by guiding its economy with national plans. These national plans served, as stage theory states, to rationalize the stages of development. For Nigeria, as elsewhere in the Third World, it is an unresolved question whether the best planning strategy is to balance the stimuli given to market forces or to funnel government resources to the needs of a particular sector. Nigeria chose the latter course, although it is not clear whether any other choice could have been implemented, given the poor state of information about Nigeria's economy and the meager infrastructure it had to work with.

As discussed in Chapter 3, the emphasis on trade and capital accumulation as fuels for development has been called into question, and the ideas of the theorists who are pessimistic about the ideas advocated by neoclassical theorists are laid out. In this way, it becomes clear that Nigeria surely had several perspectives from which to choose in her quest for autonomous development.

3 Critical Development Theory and Nigerian Development Strategy

INTRODUCTION

In Chapter 2, the classical and neoclassical perspectives that Nigeria followed in its pursuit of development were discussed. Its strategy of development was said to be derived from theories written to explain the experience of developed economies. This chapter presents the body of criticism that has challenged neoclassical development ideas, with particular focus on the ideas of Third World theorists who see development and underdevelopment from a structural point of view. Certain criticisms of neoclassicism were recognized and accommodated by Nigerian planners. However, those criticisms concerning the structural relationship between development and underdevelopment were not acted upon. By examining the criticisms of neoclassicism, especially those dealing with structural issues, one can learn about features of the development process that Nigeria's leadership either could not or would not recognize. The failure of Nigeria to respond to the structural problems it faced after independence (discussed in Chapters 6–8) can be understood as at least in part a result of its interpretation of the development process via the neoclassical paradigm.

KEYNESIAN DEVELOPMENT PLANNING

Keynes, who was a policy maker in England, wrote *The General Theory of Employment, Interest and Money* (1936) at a time when the reduced output and massive unemployment of the Great Depression was plaguing the Western world. His recommendation for deliberate state action to take care of any recession or depression—even at the risk of sacrificing the balanced-budget concept—was meant to be a corrective to the laissez-faire attitude of neoclassical development theory.

Keynes believed that capitalism is a mechanism that can be repaired and improved so that it can contribute to, rather than obstruct, economic development. However, he forecast stagnation as the likely condition of modern capitalism and agreed substantially with Marx's "breakdown" theory that capitalism would suffer from chronic underconsumption, general overproduction, and a secularly declining rate of profit (Tsuru 1954). Keynes proposed that many crises could be avoided only by abandoning laissez-faire capitalism through deliberate state action. His message was that market economies do not, in their ordinary functioning, necessarily gravitate toward full employment growth. In this respect, his *General Theory* represents a structural criticism of capitalism.

Keynes identified three new tools of analysis as the basis for developing the necessary state actions to forestall stagnation: the consumption function, the investment-demand function, and the liquidity preference function. The consumption function relates consumption to income ($C = f(Y)$) and tells us that consumption increases at a lower rate than income—or, in his terms, the marginal propensity to consume (that is, change in consumption as a fraction of the change in income) is always smaller than unity. This is because people have the habit of saving part of their income (unless it is very low) and a larger proportion of any addition to net income is saved. This gap between consumption and income, representing the community's savings, means that supply (income) does not create a proportional demand. Hence, the ability to market total potential output will depend on the willingness of business firms to invest the amount saved in the production of capital goods in order to increase future production of consumer goods.

However, the motivations of business firms to invest are entirely different from those of the savers. They are also concerned with the cost of investing these savings, that is, the interest rate. Hence the investment-demand function ($I = f(Y, i)$), which relates the rate of investment to the marginal efficiency of "capital in general which that rate of investment will establish" (Keynes 1936:136). (Note that the rate is an aggregate of the marginal efficiency schedules, of different types of capital goods.) The marginal efficiency of capital, then, is the ratio between the expected yield from one additional capital asset and its supply price. Therefore, as the investment in any capital good increases, its marginal efficiency will decrease.

Keynes's third tool of analysis—the liquidity preference function—relates people's desire to hold cash (rather than lend their savings) to the rate of interest: $M = L(Y, r)$ (Keynes 1936:19). People's demand for cash may have one of three motives: transactions, precautionary, and speculative. It is the last motive that is of interest here. The first two are not influenced by the rate of interest, but by the level of income, and therefore do not directly affect investment decisions. The higher the rate of interest, the less people will tend to hold on to their savings in the form of cash, and vice versa—the liquidity preference schedule is a declining function of the rate of interest. It is evident how the monetary authorities can affect the liquidity preference function by means of the tools of monetary policy. Keynes's contribution to the theory of saving is his observation that people save without having any particular intention to invest, and may not want to invest their savings—in which case such saving becomes a leakage out of the income stream, in the sense that it does not contribute to furthering the production process.

Keynes also contributed to the idea and establishment of the International Bank for Reconstruction and Development (World Bank). Here, Keynes contradicted Marx and his followers, who considered international heterogeneity in the economic field as a permanent way in which advanced capitalistic nations exploited backward nations (Heimann 1952). Keynes advocated pooling world resources to benefit poor nations. This challenged the Marxian view that the advanced nations would resist such policies in order to maintain domination and "center-periphery" polarization. Keynes looked upon international economic homogenization as a path to universal prosperity and lasting world peace. He believed the World Bank to be the proper instrument for making the saving propensities of the world's developed nations compatible with the development needs of the poorer nations. The novel idea of a World Bank for the specific purpose of reconstruction and development owes much to the imagination and internationalism of Keynes.

Keynes's idea that government planned expenditures could stimulate investments was key to revising the developed world's understanding of the basis for sustained development. No longer would it be appropriate to rely on market forces to establish the basic rate of growth or the composition of growth. The various national development plans in Nigeria, as discussed in Chapters 5–8, reflect this idea.

HARROD-DOMAR DYNAMIC GROWTH MODELS

The recognition and understanding of the fundamental relationships between capital accumulation and economic growth in a modern dynamic (and advanced) economy were initiated by Sir Roy F. Harrod in "An Essay in Dynamic Theory" (1939). Starting from a Keynesian framework, he developed his dynamic theory to explain how steady growth occurs over time and how it can deviate from its equilibrium path.

Arguing that investment depends on the change in the level of effective demand, he developed the acceleration principle as a theory of investment, thus making his (induced) investment model more dynamic than Keynes's (autonomous) investment model.

Prof. Evsey Domar developed his theory of growth in articles published in the *American Economic Review* in 1947, reprinted in his *Essays in the Theory of Economic Growth* in 1957. Like Harrod, the type of economy he had in mind was one with a high saving ratio and a high productivity of capital. But, unlike Harrod, he focuses not on what investment would be, but on what it should be to sustain the conditions needed for the maintenance of full employment over a period of time—or, more exactly, the rate of growth of national income that the maintenance of full employment requires (Domar 1957:84).

The Keynesian view of the role of capital in economic growth, which forms the basis for the Harrod-Domar model, states that an increase in the capital stock (net investment or net capital formation) is necessary for an increase in the income flow to occur in the future. But investment cannot occur without savings—saving being a sacrifice of current consumption. If all savings are invested, the increase in the future consumption that can be expected is provided by the relationship between new capital and new output (or incremental capital output ratio, ICOR). Thus, the increase in the flow of future production is equal to the increase in the capital stock divided by ICOR. The actual limit to investment is, therefore, the sacrifice of current consumption that society is willing to bear—the propensity to save.

The Harrod-Domar model was developed to affect this propensity. The Keynesian framework sets as the condition for national income equilibrium the equality between desired saving and desired investment. In order for full employment to materialize, we must have equality between aggregate spending and potential output. A net investment greater than zero is believed to bring a greater future potential output—capital formation means a greater flow of output. Also, because the current level of aggregate demand will be inadequate to guarantee future full-employment equilibrium, certain adjustments are required.

In order to be able to utilize all future capacity and maintain equilibrium, the rate of growth of the economy is determined by the following relationships:

Representing all changes by d,
national income by Y,
savings by S,
propensity to save by s,
net investment by I,

ICOR by k,
and the growth rate of income by g,

we have the following equations:

$$S = s \cdot Y$$
$$I = k \cdot dY.$$

Since $S = I$ at equilibrium, then
$$k \cdot dY = s \cdot Y$$
$$k \cdot dY/Y = s$$
$$dY/Y = s/k$$
$$g = s/k.$$

That is, the growth rate of income is equal to the propensity to save in the economy divided by ICOR. For instance, with a saving rate of 10 percent of national income and an ICOR of 2 (that is, one unit of new output per two units of new capital), the rate of growth would be 5 percent. If the rate of growth diverges from this "warranted" 5 percent, an inflationary or recessionary movement may set in (Harrod 1963; Domar 1957).

The Harrod-Domar growth models are designed to indicate the conditions of progressive equilibrium for an advanced economy, and therefore do not directly address the conditions prevailing in a developing country such as Nigeria. Nevertheless, they are important, not only because they represent a stimulating attempt to dynamize economic development theory but also because they are capable of being modified so as to introduce fiscal-policy parameters as explicit variables in the economic growth of a developing country.[1]

Specifically, a developing economy can treat Harrod's autonomous investment ratio (k) as a key factor in development planning. In the Nigerian situation, autonomous investments by the government were concentrated in social overhead capital projects with the hope that this would stimulate investment in the productive sectors. The results of the strategy are discussed in Chapters 6–8.

SOCIAL DEVELOPMENT: A CRITIQUE OF THE CLASSICAL AND NEOCLASSICAL VIEWS

After World War II, a new field of economics, concerned specifically with the economic development of backward countries, began to take shape. It was aligned with Western socialism and stressed distribution as part of the development process. It reexamined the relationship between international trade and economic development. Among the scholars who have contributed to this field are Gunnar Myrdal, Hans Singer, and Hla Myint.

Generally, the views of these development theorists conflict with the classical and neoclassical perspectives. As discussed earlier, the classical and neoclassical economists believe that any country should experience economic growth as its export sector expands. But the social development economists argue that not every type of potentially leading sector is successful in spreading its growth to the rest of the economy, particularly in the developing countries, which must rely on raw materials sectors for trade. This view is true in the case of the petroleum sector of Nigeria, as shown in Chapters 6 and 7. These economists maintain that primary exports, which include food and beverages, petroleum, lumber, and agricultural and industrial raw materials, should play a secondary role in an economic development drive.

There are many reasons cited for the failure of primary export products to transmit an expansion to the rest of the economy in a developing country such as Nigeria. First, the operation of the "terms of trade" tends to reduce the purchasing power of a unit volume of primary products exported by developing countries. This creates a harmful and unfavorable situation for the developing nation by increasing its import bills at a faster pace than its export earnings.

Second, the "factors of production" argument stresses the point that the consequences of factor inflows into developing countries have not been favorable to these nations (Myrdal 1956). It is believed that most of the factor inflows resulted from the needs of primary production for export, and that production for export was organized like an enclave with more external than domestic contact. Hence, the skills and technical know-how that were developed in the export sector did not spread to the rest of the economy. Furthermore, most of these enclaves are owned by foreign investors, enjoy a monopsony position in the market for labor and other domestic goods and services, and have a monopoly within the country over the goods produced and sold. According to Singer (1950), the foreign investors concentrated on the development of the country's natural resources for export, to the neglect of production in the domestic sector. These enclaves thereby created a dualistic and lopsided economic structure that is economically inefficient and deters the development of the backward sector of the economy. Primary production for export in Nigeria and other developing countries, therefore, cannot be relied upon to stimulate the development of the rest of the economy because there are few spillover effects emerging from enclaves. The situation described above is the same whether the export activities are in the hands of foreign investors or domestic capitalists.

There is also the development deterrent operation of the international "demonstration effect" (Duesenberry 1949; Nurkse 1953). Through international trade and contacts, the developing countries have been integrated into the international economy. The concomitant result is that the developing nations tend to imitate the consumption patterns of the developed nations, thus raising their already high propensities to import. Hence, developing nations experience difficulties in increasing domestic savings, and further exacerbate their balance of payments problems. The Nigeria situation is covered in detail in Chapters 5–7. Therefore, to control this "demonstration effect," the state has to abandon laissez-faire and must play a more active role in controlling imports. But, as discussed later, import control could prove to be problematic and is likely to backfire.

Gunnar Myrdal, who conceived the "cumulative mechanism of causation"—a prelude to dependency theory—has argued that as a mechanism international trade has, by its very operation, led underdeveloped countries to stagnation or impoverishment, and developed countries into automatic cumulative growth:

> Contrary to what the equilibrium theory of international trade would seem to suggest, the play of the market forces does not work towards equality in the remunerations to factors of production and, consequently, in incomes. If left to take its own course, economic development is a process of circular and cumulative causation which tends to award its favors to those who are already well endowed and even to thwart the efforts of those who happen to live in regions that are lagging behind. The backsetting effects of economic expansion in other regions dominate the more powerfully, the poorer a country is. (Myrdal 1956:47).

According to this argument, international trade and contacts have favorable and unfavorable effects. The favorable or "spread" effects are believed to be weaker than the unfavorable or "backwash" effects (Myrdal 1957; Higgins 1968). In addition, Singer (1950) felt that domestic industry, as distinct from production for export, can effectively promote economic progress in developing countries because of certain intrinsic traits that primary production for export does not possess.

Before concluding the debate on international trade, it is important to observe that the early visions of this activity as an engine of growth have given way to decidedly less optimistic views of its importance for development. There are several critiques of international trade relations that, though powerful, do not suggest a general perspective on development. Yet, they suggest a growing sense within traditional theory of its failure to explain development, particularly in a post-colonial context.

For instance, Ragnar Nurkse (1961) is particularly pessimistic about the effect of international trade on the economies of developing countries. He presents three basic arguments against the traditional positive view. First, that in developed countries, industrial efficiency is rising through increasingly effective economizing on inputs of raw materials. That is, raw input materials from developing nations are systematically being replaced by synthetic substitutes and recycled waste metals and other materials. This has resulted in depressed demand for natural rubber, cotton, hides, skins, and silk. Thus, to the extent that a developing country's trade is based on domestic raw materials industries, trade may be a poor strategy for encouraging development. In his second argument, Nurkse stresses that agricultural protectionism in advanced countries is a bar to entry of products from much of the Third World, especially the tropical countries. Again, development strategies based on trade would seem doomed if the developing country is offering primarily agricultural goods. Finally, Nurkse observes that farm products often suffer from low income elasticity of demand. That is, a rise in the real incomes of the developed nations is accompanied by a falling share of farm products. Thus, demand for Third World produce can be expected to fall over the long term, jeopardizing development based on the most common developing country export—agricultural goods.

Developing countries can expect to find unfavorable terms of trade generally, and should not hope for significant internal economic development as a result of expanded trade. The gains from trade depend on both the volume and the terms of trade. The greater the volume of trade, the greater the average gains from trade for each country through international specialization; and the more favorable the terms of trade for an exporting country, the greater the marginal gains to that country (Meier 1980). The more favorable a country's terms of trade, the less it must export for the same volume of imports. In other words, it can import more for the same volume of exports. Therefore, as its terms of trade improve, there is a rise in its income from a given output. But the terms of trade are determined by the intensity and elasticity of each country's demand for imports, within a region of potential gain between the trading countries' domestic exchange ratios. The country with less intense and more elastic demand for imports—usually a developed country—will enjoy the more favorable terms of trade as the international price line approaches the other country's domestic cost ratio (Meier 1980).

In his study of commodity terms of trade (omitting services), Kindleberger (1956:239) found empirically that "in the European context, the terms of trade favor the developed and run against the underdeveloped countries." He also found evidence that the developing nations are less flexible than the developed nations in their capacity to shift resources out of products whose prices are falling and into products whose prices are rising. The exporting-developing nations are subject to collusion and monopsonistic buying practices among the importers in developed countries, and to monopolistic pricing practices from sellers in developed countries.

In addition to the critique of trade as basis for growth, the social development theorists question whether other conditions assumed by traditional development theorists are appropriate when talking of Third World development. Conventional theorists assume that developing countries should import technology, but fail to consider the difficulties faced by these countries in adapting to Western technology. Advanced technology has evolved along lines suited to the conditions of developed economies. It uses relatively little labor and a great deal of capital, and its operation requires skilled labor and technically trained personnel. On the other hand, in a developing country, capital is very scarce, labor is abundant, and there tends to be an acute shortage of technical and managerial staff. Importation of advanced technology into developing countries could, in this respect, be likened to transplanting the fruits of development rather than planting its seeds.

Convention development theorists view rapid population growth as an obstacle to development. Emphasis tends to be placed on the rate of increase of population and on the population density relative to land and other resources, and on migration from rural to urban areas, as well as from poorer to "richer" developing countries. The question of the appropriate utilization of the increased human resources in developing countries, particularly in ways that could make them self-sufficient in domestic food production, tends to be ignored. (The issue of rural-urban migration in Nigeria is dealt with in Chapter 8.)

Even though neoclassicism would prescribe trade and capital accumulation for development in the Third World, what is often overlooked is that modern developing nations face a different, and generally less favorable, international environment than did the industrially developing nations of the 18th and 19th centuries. Social development theory, therefore, is a structural paradigm that considers the relationship between sociopolitical and economic structures. Also, it recognizes the role of income distribution in economic development. It points out the weaknesses of neoclassicism and rejects the neoclassical paradigm. What was true in the 18th century is not necessarily true today.

As evidenced by the Nigerian case, there may be unfavorable effects from a country's dependence on exports. A negative acceleration effect, resulting from negative growth rate of exports over the previous period, can lead to a reduction in the economic activities of producing sectors. The concomitant decline in foreign reserves can lead to a sharp reduction in the importation of capital and consumption goods, thus limiting domestic growth. Furthermore, export demands may mislead countries into deadend specializations with no favorable linkage effects and with a high probability of sharply falling prices in prospect; they can also lead a country to neglect the development of more productive domestic-oriented industries. Subsidizing domestic prices of exports in order to entice foreign buyers may result in allocation of scarce resources toward the importing countries.

The above represents a condensed expression of the position taken by social development economists. In brief, it is the conclusion of this group that a primary goods export sector, or a particular line within that sector, cannot fulfill the role of leading sector in a developing country with a trade pattern based on raw material exports.

Despite all these criticisms of international trade as an engine of growth, Nigeria, as is discussed in later chapters, chose to emphasize export trade as the key to development while increasing state ownership of productive resources.

THIRD WORLD DEVELOPMENT THEORY

Partly provoked by the failures of traditional theory, and partly the result of Third World theorists seeking to explain development from the historical context and experiences of the developing world, a new body of development theory has been created that is sharply at odds with the conventional paradigm. There is a wide variety of Third World development theories, some traditional Marxist, some revised Marxist, some without a dominant Marxism but with a nationalist focus. Yet, they agree on certain themes: that trade is organized to benefit developed countries; that past development was colonial in nature and, as such, cannot be used as a model for the Third World; that internal social and economic factors should take precedence over both the speed of development and linkages with the developed economies; and that development in the post-colonial context is constrained by the economic conditions of the developed world—that the developing world is struggling against an international economic order that forces its dependence for development on the advanced economies. In line with the above, Dos Santos (1970:231) defines dependency as follows:

> A situation in which the economy of a certain group of countries is conditioned by the development and expansion of another economy to which the former is submitted. The relation of interdependency between two or more economies, and between these and world commerce, assumes the form of dependency when some countries are able to expand and self-propel themselves, while the other countries can only do so as a result of that expansion, which can act positively or negatively on its im-

mediate development. Either way, the basic question of dependency leads to a global situation in the dependent countries whereby they are placed in a backward position in relation to, and under the exploitation of the dominant countries.

Lester Pearson (1970:17), in describing the environment in which dependent countries exist, states:

We should never forget, in short, that the developing people do not start from scratch in a new world but have to change and grow and develop within a context unfavorable to them, because in the past their position has been so largely determined by the interest of other nations. If we forget this historical context, we will not understand the problems that now exist; nor will development cooperation to solve them be likely to succeed.

The two major theoretical frameworks that embody the ideas and analytical approaches outlined above will be discussed in detail: dependency theory and the internationalization of capital framework.

Dependency Theory

The concern with underdevelopment and dependency evolved from the earlier work on imperialism and concerns its effects. Relationships of dependency and descriptions of exploitation and deformation were prominent in the writings of Marx, Lenin, and Trotsky.

In his first volume of *Capital*, Marx presaged the dependency school's idea of development of underdevelopment:

A new and international division of labor, a division suited to the requirements of its chief centers of modern industry, springs up and converts one part of the globe into a chiefly agricultural field of production, for supplying the other part which remains a chiefly industrial field. (Marx 1967:I; 451)

In his views, merchant's capital

functions only as an agent of productive capital. . . . Whenever merchant's capital still predominates we find backward nations. This is true even within one and the same counry. . . . (Marx 1967:III; 372)

In his "Imperialism: The Highest State of Capitalism," Lenin referred to the idea of dependent nations on the periphery when he argued thus:

Not only are there two main groups of countries, those owning countries, and the colonies themselves, but also the diverse forms of dependent countries which, politically, are independent, but in fact are emeshed in the net of financial and diplomatic dependency. . . . (Lenin 1967:I; 742–43)

Lenin thus provides the first systematic analysis of capitalist development in backward nations and what would be the core of the dependency analysis (Cardoso 1972).

Twentieth century Third World theorists, while agreeing with the projected results of early Marxist theory, disagreed with the view of imperialism because it ignored the internal structure of underdeveloped nations. Their writings tended to shift analysis from exclusively external factors and dynamics to internal considerations and an analysis of economic and political conditions more appropriate to Third World conditions than those provided in conventional class analysis.

The main focus of dependency theory is that underdevelopment has causes external to underdeveloped areas. Some of the main propositions advanced by dependency theorists are the "center" or "core/periphery" theory, the theory of unequal exchange, and the dualism critique.

The core or center is made up of the advanced economies, and the periphery represents the developing countries. This notion is also used to describe the internal structure of nations—where one part of the country exists as a colony of the dominant center. The second proposition is the critique of dualism. It directs attention to imperialism's role in national oppression and underdevelopment. It disagrees with the notion that the reason for underdevelopment can be found in the social and cultural characteristics of the traditional and backward sectors of developing countries. Instead, dependency theorists blame the historical underdevelopment of Third World countries on colonialists and the few wealthy countries that dominate the world markets. The theory of unequal exchange is a critique of the Ricardian doctrine of comparative advantage. This theory stresses that underdevelopment is related to the adverse conditions that peripheral countries face in the world market (Emmanuel 1972b).

After World War II, nationalist sentiment in the periphery was accompanied by an outcry aginst imperialism, the demand that national resources be preserved, and an insistence that the domestic economy be transformed through state-guided national capitalism. For example, Nkrumah's Ghana, Nasser's Egypt, and post-independence Nigeria argued for national economic and social autonomy. These countries, which had recently attained independence, showed a pre-independence heritage of colonialism that focused on exploitation of natural resources and depended on cheap labor.

These ideas were also expressed by Raul Prebisch, who in his writing divided the world into two parts: a center consisting of industrialized countries, and a periphery made up of underdeveloped countries. He argued that because of distortions in international commodity markets, there has been a deterioration in the terms of trade for the periphery. The resultant decline in foreign reserves impedes the development of Third World nations. Prebisch believed that the state must play a major role in coordinating private and public enterprise, and in providing intervention in the form of subsidies, tariff protection, and import substitution that would allow the periphery to counter the dominance of the center and move toward an autonomous capitalist solution (Prebisch 1950). The same views were expressed by the Chilean economist Osvaldo Sunkel, who believed that dependency links the development of the international capitalist system to the local processes of development and underdevelopment within a society. He saw underdevelopment and development as simultaneous processes representing two facets of the evolution of capitalism (Sunkel 1972).

The Mexican political sociologist Pablo Gonzalez Casanova presented a different version of the autonomous capital approach. He believed that imperialism was somewhat controlled in Mexico but that an internal colonialism similar to the colonialist relationship between nations was present. A society ruled by Mexico City dominated and exploited a marginal society of Indians. This type of dualist relationship created a situation in which the Indians became dependent on Mexico City. The solution would be a democratic revolution and a class alliance of the masses with the bourgeoisie to oppose imperialism and to support capitalist democracy and peaceful development (Gonzalez Casanova 1970).

A quite different view, which is revolutionary in outlook and oriented toward socialism, was also evident in post-World War II writings of Third World intellectuals. It opposed imperialism and saw capitalism as a negative force in the periphery. Frondizi (1954) identified two imperialisms: British commercial imperialism and U.S. industrial imperialism. He related imperialism to the emergence of a national bourgeoisie in Argentina, and further argued that capitalism in its various forms was responsible for underdevelopment and dependency on world capitalism (Frondizi 1954; 1957). Prado (1967) argued that the Brazilian economy passes through a cyclical pattern of prosperity and decline as a result of its dependence on international trade and on Portugal. According to him, socialism achieved through revolutionary means would be the only way to establish Brazilian economic autonomy (Prado 1966).

Frank (1966; 1967) also took a revolutionary stand, arguing that capitalism, rather than feudalism, has been dominant in Latin America since the colonial era. He attacked the notion that the national bourgeoisie is a progressive force for change. He referred to the three capitalist contradictions of surplus expropriation/appropriation, metropolis-satellite polarization, and continuity in change. Frank (1967) developed the idea of capitalist development of underdevelopment whereby each phase of imperialism is a further development of underdevelopment in continuing the appropriation of surplus from the satellite to the metropolis. He refuted the notion that a diffusion of capital and technology from the advanced capitalist countries to the less developed countries leads to the destruction of feudal stagnation and the rise of universal capitalist development, prosperity, and democracy. Instead, Frank observed a flow of resources from the less developed to the more developed nations. He later moved from a theory of underdevelopment and dependency to analyses of the world captalist systems, in which he examined past and present world economic crises. He discussed the prospects for autonomous socialist development, which involves unlinking from dependent capitalist development that has shaped the processes of development and underdevelopment in many countries.

Frank saw a single world economic system of many unequal,parts: a developed North and an underdeveloped South; a capitalist West and a socialist East. Also within the world system, there are pre-capitalist, capitalist, and post-capitalist relations of production (Frank 1981; 1983). He stressed that external unlinking and internal participation must be combined with social and political mobilization to bring about rapid structural change (Frank 1983). Immanuel Wallerstein has written extensively on the world economic system (1974a; 1974b; 1974c; 1976; 1978; 1979; 1983; 1984).

The neocolonial dependence model attributes the existence and perpetuation of developing nations' underdevelopment to the historical evolution of a highly unequal international capitalist system (Todaro 1977:55). It also assumes that an international system dominated by the "center" or the advanced economies has made it virtually impossible for the "periphery" or the developing economies to be self-reliant and independent in their development efforts.

Samir Amin (1974b) described a fundamental difference between capital accumulation and economic and social development characteristic of a self-centered system and that of a peripheral system. The main features of the center are mass consumption and capital goods, while those of the peripheral (and dependent) nations are export and consumption of luxury goods. To Amin, dependence is a complex set of phenomena related to the accumulation process of the capitalist mode of production in its monopoly and imperialist phase. The peripheral economy came into existence as a result of demands imposed on it from the outside: the supply of primary goods. In the colonial and the post-colonial eras, most developing nations, including Nigeria, were dependent mainly on the export of agricultural and mined raw materials for foreign exchange and the bulk of their national income. With the revenue obtained from the exports, these developing nations imported consumer goods, including food, from the advanced countries. The consumption pattern that followed this development of an export economy was, therefore, consumer-oriented and generated by import-substitution industrialization. This meant, and to a large extent still means, production for the rich few, the upper middle class. It left the majority of the population out of the stream of development. Therefore, however large the wealth of the nation, this sort of development could rightly be termed "growth without development" (Clower et al. 1966), because it leaves the masses empty-handed.

In short, underdevelopment, according to this model, is not the original state of Third World countries, but the product of structural dependence incorporated into the capitalist system. Continued development within this structure has led to a polarization of wealth and poverty (Emmanuel 1979; Wallerstein 1978). Meaningful development, therefore, is a process of socio-economic and political transformation in such a manner that the masses are meaningfully participants in, as well as sharers of, its costs and benefits.

Internationalization of Capital

The internationalization of capital is a relatively new approach to the study of the world economy and to the problem of dynamics of global capitalism and international economic relations. It explains the effects of the growth of the world economy on the productive structure of each of the participating nation-states.

David Barkin (1982) has argued that the "internationalization of capital" is a much more fruitful approach to studying the world economy than is dependency theory. He has suggested that, rather than trying to explain the backwardness of one part of the world vis-à-vis other parts, political economists should focus on the dynamics of the emergence of the international capitalist economy.

The basis for this theory is the intertwining of Marx's three circuits of capital: money capital, productive capital, and commodity capital (Marx 1967:II). The early importance of commercial capital in the form of foreign trade has been noted. Lenin also emphasized the importance of money capital in his essay on imperialism. But in recent years there has been an expansion of productive capital on a global scale whereby the various facets of capitalist economic life have been integrated into an increasingly unified global market.

As the multinational corporations (MNCs) emerge as the main organizers of production on a global scale, the dynamics of the world capitalist economy cannot be understood with reference to a single nation or group of nations. Rather than focusing on international trade (circulation) as the key element to be analyzed, the "internationalization of capital" approach examines the determinants of production on a global scale. It raises questions about the new international division of labor. Rather than descriptively stratifying production into primary, intermediate, and tertiary, this perspective distinguishes between those aspects of the production systems that contribute to the reproduction of the labor force (means of subsistence or wage goods) and those that are directed toward capital accumulation.

The emphasis of internationalization of capital is on the problem of the relationship between global tendencies and MNCs, on the one hand, and the individual nation-state, on the other. In this way, the MNCs are shown to be actors in the process of the accumulation of capital on a global scale. This school also argues that national policies are influenced by international events and pressures, and that the eagerness of a country to integrate its domestic economy with those of its trading partners, while a product of the conviction of national policy makers that this is in the country's best interest, actually contributes to further internationalization of the domestic economy by increasing the impacts that allocative decisions in the international economy have on a country's internal socioeconomic structure. Barkin writes:

> Obviously production still has a geographical specificity, but this approach argues that the logic of allocative decisions is now global even when transnational capital is not involved in a particular activity. That is, even when investment and production decisions are made by national governments or local capitalists, it seems increasingly clear that global economic and political structures strongly influence the individual decision maker. It is possible to state this even more strongly: international markets and economic power structures are increasingly determining the individual decisions made in ever more isolated parts of national economies, even when "noncapitalist" productive groups are involved, such as peasant producers in many Third World economies. (Barkin 1982:158)

Also, some proponents of the internationalization of capital perspective have recognized that dynamic elements exist in the changing historical conditions for capital accumulation in the Western countries, particularly since the petroleum price increases of 1973. European capital is increasingly being directed toward certain countries of the Third World, with the intention of finding new areas of investment and new markets. This has resulted in an increased differentiation among the countries of the Third World. While most still suffer from lack of development, a few are believed to be in the process of establishing a basis for national capital accumulation. According to this group, the state is being used by the dominant classes in the Third World to transform the economy in order to establish internal relations of production. This is done by establishing the material preconditions for indigenous capitalist production (infrastructure, investment codes, and legal frameworks), and by extracting economic surpluses primarily from the agricultural sector, presumably to be reinvested in agriculture and industry (Marcussen and Torp 1982). The state has thus taken an active, economically interventionist role.

The character of private direct foreign investment also has changed. Instead of foreign control over management, what have become prevalent are participation in management, technical agreements, loansd, production sharing, and supply contracts. In addition, after 1973, importation of a large-scale producer goods complexes, often in the form of turnkey projects, increased. Marcussen and Torp (1982:11) write:

> This introduction into the Third World of large, highly developed and technologically sophisticated production complexes has taken place, as has the growth in private bank lending, particularly in the years of world economic crisis provoked by the increase in oil prices (and often financed by the huge oil revenues piling up in the Western banks). Already these new investments are reflected in the changing export pattern of some peripheral countries. At the aggregate level, the periphery's export statistics are still dominated by raw materials, agricultural as well as mineral. But for these Third World countries with a development strategy based on creating "sound investment climates" (inter alia, low wage levels, infrastructure built by the state, tax and custom duties exemptions) the export of finished or semi-finished manufactured goods to the Western world has been increasing drastically.

However, Amin (1977) called this another phase of imperialism and said that it could not lead to a self-reliant economy. Therefore, proponents suggest that instead of proposals for a code of ethics for corporate behavior or new rules for negotiatng with the MNCs, what is necessary to halt internationalization is a new strategy for producing the goods that the masses of people need for their basic survival rather than concentrating on production for export. The focus should be on changes in the productive apparatus as part of a longer-time political struggle for structural transformation (Barkin 1982:161).

Summary of Third World Development Theory

Dependency theory was conceived as a response to the inadequacies of the theories of imperialism in explaining the impacts of capitalism on domestic structures of the less developed countries. While it has generated new questions and interesting analyses of Third World countries, it suffers from a lack of general and unified theory. Also, confusion over terminology often diverts investigation away from a central focus. Dependency theory has not demonstrated ways to solve the problems of capitalist exploitation. But despite these criticisms, there has been a growing interest in dependency theory throughout the Third World. Dependency theory returns us to development as national, self-sufficient, rural (or at least rural with urban), and slow. Its policy recommendations are isolationist in nature.

The internationalization of capital approach focuses on the dynamics of the emergence of the international capitalist economy. It examines the problem of the relationship between global tendencies and MNCs, on the one hand, and the individual nation-state on the other. National policies in developing nations are believed to be influenced by international events and pressures. The strategy suggested is for developing nations to change their productive apparatus, and produce the goods that the masses of people need for their basic survival rather than manufacturing for export markets. In Nigeria, this strategy was ignored, as Chapters 7 and 8 point out.

CONCLUSION

Despite the fact that trade and capital attraction principles have been called into question, many Third World countries have continued to base their development on these principles. These countries, including PPDCs such as Nigeria, have utilized both principles in their economic development efforts. The influence of capitalist development ideas on Nigerian development strategy was so strong that Nigeria never acted to significantly alter the colonial structure of its economy. Unfortunately, as Chapter 8 points out, it is not clear that even the noncapitalist development ideas could have instructed Nigeria on structural change.

Chapter 4 analyzes the various types of development planning modes available to Nigeria and the reason it chose one mode over the others.

NOTE

1. Given a target rate of growth (g^*) and an estimated k, the proportion of total resources Y to be devoted to investment (s^*) = $g^* \cdot k$.

If $g^* = 0.15$ (15 percent) and $k = 2$, $s^* = 0.15(2) = 0.30$ (30 percent) that is, s^* or MPS = 30 percent to achieve 15 percent growth.

Savings-investment gap = s* - s, that is, target MPS minus real MPS.

This is investment-limited economic growth. To cover this gap, a country may raise domestic saving, increase trade, or seek external loans/aid. The export-import gap = i* - i and relates to trade-limited economic growth. where

m = dY/di (incremental output-import ratio) that is,

m = relationship between growth and imports,

Y = national income,

i = actual ratio of imports to output (Imports/Y),

i* = g*/m.

4 Development Planning in Nigeria

INTRODUCTION

Development efforts in the Third World have for the most part been based on the assumption that two primary factors influence the movement from underdevelopment to development—a country's capacity to extend international trade and to attract capital. Responding to the mounting pressure for development, leaders in the Third World have turned to national planning in order to build and operationalize strategies for economic and sociopolitical development. Planning, it is believed, results in blueprints for future development; the specification of courses of action for the achievement of desired goals. National planning is seen by the developing world as a conscious effort of a central government to influence, direct, and control changes in the principal economic sectors of a country in order to establish the process of internally directed development. This chapter examines Nigeria's planning philosophy, model, and organization that were intended to bring about autonomous development.

THE ATTRACTIVENESS OF DEVELOPMENT PLANNING

There are two distinct reasons why, throughout the developing world, the quest for rapid economic progress has been predicated largely upon the formulation and implementation of comprehensive national plans. They concern nationalist ideology and the association of planning with modernism (Olayiwola 1985:122).

The Ideology of Nationalism and the Rise of Planning

Patrick McAuslan (1980:xii) provides an explicit definition of ideology:

> Philosophy denotes a carefully prepared and thought out set of values and ideas. Ideology, on the other hand, denotes values, attitudes, assumptions, hidden inarticulate premises that may not be well thought out and are usually disguised rather than spoken out loud. Ideologies must be pieced together and drawn out from a mass of writings and statements official and unofficial, ostensibly dealing with matters of substance—rather than the world of ideas and beliefs.

As an ideology, nationalism implies full independence and that, once won, independence should be used to benefit the nation. This ideology has been a strong and critical force responsible for anticolonial movements, particularly in Africa. For example, it was the concerted pressure of nationalist movements that led to the independence of Nigeria in 1960.

Many developing countries in post-colonial settings have tended to use nationalist sentiments to build support for policies that would drastically reduce the domination of their economies by foreigners. Nigeria, for example, by means of its Indigenisation Decrees of 1972 and 1977, followed this pattern by indigenizing many of the major industries operating in the country. The nationalist assumption is that economic resources are most beneficially employed by domestic concerns. The less that goes to foreigners, so the reasoning goes, the more that remains for the indigenous people.

The use of nationalism to fortify political regimes has great impacts on the effectiveness of economic planning. In many cases, economic planning tends to be a form of power. Political leaders often see economic policy and planning as part of the means to establish and sustain national control. For them, economic development may be a by-product of more important political goals rather than something to be achieved for its own sake. This lack of devotion to economic development is manifest in the way plans reward supporters, keep prices artificially low to prevent discontent, and fail to adhere to strict economic measures when these are called for. Nationalism therefore, can work both against and for policies conducive to economic development, depending upon whether there is coincidence between the political interest of leaders and economic development policies. This raises the question of whether economic development has been the central concern of political leaders. Nationalism tends to be used in rationalizing national planning as a tool that will ultimately bring about indigenous economic development. Hence, nationalism is a common ideology in the developing world.

National Planning and Rationalist Symbol

The second reason why planning is attractive to developing nations is that it is seen as imparting scientific objectivity to the development process; that is, planning serves to bring logic and rationalization to development. Ministries of Economic Planning in developing countries are regularly engaged in the process of drawing up development plans in order to set forth in a logical and consistent manner the priorities, goals, and aspirations of their governments. For these countries, planning involves a belief that economic development can be determined by analysis and human action based on reasoning; and that the efficiency of development projects can be improved by systematically reviewing both the policies and the investment projects as parts of a whole, thereby forming consistent and complementary development packages.[1] Planning is often used as if it is equivalent to rational action. The assumption is that once the norms associated with rationality—efficiency, consistency, coordination—are followed, better decisions will result. There is also the concept of synergism, which suggests that complementary and integrated action programs are more effective and more efficient than the sum of the results of these actions taken individually.

Planning is also believed to be systematic, which implies that planning is orderly, knowing the right variables and the correct order in which to include them being among the factors in making decisions. Developing such systems and models for national development in the Third World is viewed as modern. A national plan becomes the stylish thing that every developing nation has to have in order to show that it is modern and progressive, and to impress the owners of foreign capital.

National planning often emphasizes economic self-sufficiency, and for this reason, as an instrument it often becomes allied with the notion of modernity and industrialization. Leaders and planners in developing countries tend to think that the discovery and exploitation of improved methods of producing wealth, primarily in manufacturing, but to an increasing extent in the extractive industries, is what is meant by modernity. What is least recognized is that, as in the case of Nigeria, industrialization can actually create a kind of "subsistence urbanization" (Breese 1966:5, 99; Damachi 1972:52), of individuals living under conditions that may be even worse than in the rural areas from which they have come without the kinds of skills or the means of support that will permit them to do more than merely survive. Modernity means more than industrialization. It involves solutions to the associated problems of political stability, relative economic stability, decent living standards, and an orderly change in social structure.

NATIONAL PLANNING IN THE WORLD ECONOMIC SYSTEMS

In most Third World countries, national planning was installed by colonial powers as a mix of privately and publicly controlled activities. The guiding principles for their planning systems derive from Western economic ideals, typically expressed in terms of capitalist versus socialist development.

Planning in Capitalist and Socialist Economic Systems

In capitalist economies like those of the United States, the United Kingdom, and Japan, planning tends to consist of efforts by the national government to attain rapid economic growth with high employment and stable prices through various instruments, such as monetary and fiscal policies, and international trade relations. In the United States, national policies are formulated by the Federal Reserve Board, the Council of Economic Advisers, the Office of Management and Budget, and congressional committees whose task is to find goals and policies that embody them. The policy instruments are active but indirect (Todaro 1971). They are active in the sense that they steer the economy in a desired direction. They are indirect because they are intended to create favorable conditions that will influence private decision makers to execute plans that are conducive to stable economic growth and profit maximization. In most capitalist economies, no detailed economic plan in the form of a set of specific targets is drawn up.

Planning in the Soviet Union did not begin until 1928, and its originator was not Lenin but Stalin (Petersen 1966). In the Soviet-type economies of eastern Europe and Asia, the government actively and directly controls the movements of the economy through a centralized decision-making process. A complete and comprehensive national economic plan is drawn up on the basis of a specific set of targets predetermined by central planners. Material and financial resources are allocated in accordance with the material, labor, and capital requirements of the overall plan.

Long-term planning is inherent in the nature of socialist national economic planning. The first phase of long-term planning is a 20-year comprehensive program of scientific and technical progress (divided into 5-year periods). The next stage is a 10-year plan (in two 5-year periods), with a 5-year plan presented year by year. In other words, the whole system of socialist planning is long-range on one hand, and substantiates short-range measures on the other.

The essential difference between planning in capitalist and socialist economies is that while the former attempts to prevent the economy from straying off a desired path of stable growth by indirect but active policy instruments, the latter not only draws up a specific set of targets representing a desired course of economic progress, but also attempts to implement its plan by directly controlling the activities of practically all productive units in the national economy.

National Planning in the Third World

The type of national planning operative in the Third World is hybrid or mixed capitalist/socialist. It combines the inducement found in private enterprise economies with the control aspects of socialist economies. In mixed economies, a substantial portion of the productive resources is owned and operated by the government; the other part belongs to the private sector. The private (capitalist) sector usually consists of privately owned noninfrastructure industries in manufacturing and consumer goods (with a measure of government ownership); banking institutions (often foreign but increasingly owned by the country); the traditional subsistence sector of small-scale private farms; small and medium-size corporations in agriculture, trade, industry, and transport; and the MNCs and large-scale plantations producing for export. The public sector consists of infrastructure industry (such as roads, airports, railroads), import-substitution manufacturing, mining, large plantations, education, health, telecommunications (radio, television, telephone, telegraph), generation and distribution of electricity, housing and housing finance in urban areas, distribution of imported foodstuffs, and transportation.

Even when socialist goals were avowed, national planning in the Third World was installed to advance foreign capitalization and trade, conditions that have persisted until this day. During the colonial period, the roads that were built, the harbors that were developed, and the railroads that were built could all be linked to the transportation of raw materials from the interior to seaports for export to the colonial powers and other foreign countries. The large plantations of cocoa, rubber, coffee, palms, timber trees, cotton, and peanuts were developed for export markets. The import-substitution industries that were established were linked to international trade, in that almost all of their raw materials and other basic components were imported. In fact, colonized countries were usually limited to assembling the imported components. Furthermore, tax incentives and special privileges were granted to colonial corporations in order to attract their capital. Today, the same incentives are granted to MNCs for the purpose of attracting investments.

The Nigerian Planning Apparatus

The essential features of development planning in the Third World are influence, direction, and control. Proponents of development planning in the Third World have blended these features to yield three basic planning models: the aggregate model, the sectoral model, and the interindustry model (Todaro 1971).

The aggregate model is used to assess the maximum rates of growth of the entire economy in terms of consumption, production, investment, saving, exports, imports, population, employment, and gross domestic product. General target rates of economic development are useful starting points, but development planning usually requires a further breakdown of the plan into broad sectors of the economy whose target rates of growth can then be computed, with the objective of structural transformation of a national economic base.

The second type of planning model is the sectoral model, which consists of two different approaches to development planning. The first approach, the complete main-sector planning model, divides the economy into two or more main sectors, such as agriculture and nonagriculture, or consumption goods and investment goods. The other approach, the single-sector project model, focuses on levels of production and consumption in each individual sector, and specific development projects are planned on the basis of the potential growth of each sector. The single-sector project approach has been utilized in the preparation of the five-year plans of Nigeria and such other developing countries as Ghana, Kenya, Tanzania, and Uganda. This approach is popular in developing countries because it does not require a large and sophisticated planning organization, it serves the needs of economic nationalism, and it can be used to rationalize investment in a particular sector without having to worry about backward and forward linkages. Another reason for the popularity of this approach in developing countries is the lack of statistical data for a complete main-sector model. The disadvantage of development plans based on this approach is that they tend to suffer from a lack of internal consistency and overall feasibility. Many times, such plans become a mere collection of assorted development projects with no interconnections.

The third and most sophisticated planning model is the interindustry model. This model attempts to interrelate the activities of all productive sectors of the economy with one another. In developing economies, oriented as they are toward trade and capital attraction, and given the planned output targets for each sector of the economy, the practical utility of the interindustry model is derived from its usefulness in estimating local material, import, labor, and capital requirements, so that plans with internal consistency can be created. It is also useful in evaluating import-substitution projects. Interindustry models vary in their level of sophistication, ranging from simple input–output models to the more complicated linear-programming activity analysis models, where checks of feasibility and optimality are all built-in. The extensive use of this model in Nigeria, however, has been precluded by lack of adequate statistical data.

National planning in Nigeria dates back to the 1940s, when the British Colonial Office requested the colonies to submit development plans to facilitate the distribution of the Colonial Development and Welfare Funds. That resulted in the preparation of the Ten Year Plan of Development and Welfare 1945–1955 (see Chapter 5). From that period, national planning in Nigeria can be divided into six distinct periods: 1945–55, 1955–62, 1962–68, 1970–75, 1975–80, 1981–85. Chapters 5–8 contain the analysis of these various plans.

As a result of the World Bank mission to Nigeria in 1953, which recommended comprehensive economic planning, the National Economic Council (NEC) was established in 1955. In September 1958, the Joint Planning Committee (JPC) was established to serve in an advisory role to NEC. Other important bodies in the planning apparatus were the Economic Planning Unit of the Federal Ministry of Economic Development, and the regional Ministries of Economic Planning. The 1962–68 plan was formulated under this apparatus. In 1966, the NEC and JPC were replaced by the National Economic Planning Advisory Group. With the creation of 12 states in 1967, the Federal Ministry of Economic Development and its Economic Planning Unit took on greater roles. The Economic Planning Unit thus became the centralized agency for coordinating federal and state projects. After the civil war, this planning machinery became inadequate because of the increasing complexity of the economy. A professional planning body, the Central Planning Office (under the Federal Ministry of Economic Development), was created in 1971. It has overall responsibility for development planning in Nigeria. To advise the government on planning, the National Economic Advisory Council was created in 1972. Another important planning entity is the Joint Planning Board, which draws its members from the state and federal ministries, the Central Bank of Nigeria, and the Nigerian Institute of Social and Economic Research.[2] In order to demonstrate that plan implementation is as important as plan formulation, Plan Implementation Committees were created at the federal and state levels (Tomori and Fajana 1979).

SUMMARY

Development and national planning in the Third World, including Nigeria, have been based on trade and foreign capitalization principles. National planning in Nigeria, as elsewhere, was installed by the colonial powers as a mix of privately and publicly controlled activities, and this mixed economic system is still in place. Development planning has been used to operationalize the Nigerian strategy of development through trade and capital accumulation. Because of the nationalist characteristics of planning and because of its modernist and symbolic aspects, development planning in Nigeria has been at the center of policy making.

Nigeria's single-sector national planning apparatus is one of the most advanced planning structures in the Third World. It has a substantial British- and U.S.-trained technical and managerial class to draw upon. If an African country was to be successful in the pursuit of a capitalist development path, that country should have been Nigeria. In the 1970s, it had substantial domestic capital provided by its petroleum economy, significant foreign exchange reserves, a favorable development environment that attracted foreign capital, strong world trade relations, and a sizable industrial sector inherited at independence. The conditions for capitalist development seemed so bright that one of Nigeria's leaders during the period declared that the country would be free of savings and foreign exchange bottlenecks in its planning for growth. In light of much of the Third World's development experience up to 1970, such a declaration gives valuable insight into the optimism with which Nigeria greeted its future.

The next three chapters examine the political, social, and economic development of Nigeria from colonialism to the present. By analyzing the historical patterns of development and the major institutions that influenced development, the chapters show that Nigeria's optimism was unfounded. Its faith in petroleum-driven development overlooked the structural condition of underdevelopment embedded in its economy. As a result, Nigeria failed to realize that its petroleum economy would only reinforce its structural weakness and could not provide the basis for autonomous development.

NOTES

1. Certainly, since independence Nigeria has seen planning in these terms. The Guideposts for Second National Development Plan (FRN 1966:1), for example, declared:

Planning has been widely accepted in many parts of the world as an instrument of economic development. This is demonstrated by the existence of five year Plans in many underdeveloped countries. Substantial improvements have also been made in the technique of plan formulation.

2. For a full list of planning institutions envisaged under the Second National Development Plan, see FRN n.d.:37–38.

5   The Drive to Independence, 1860–1960

INTRODUCTION

In carrying out a meaningful analysis of the types of change that have taken place in Nigeria since independence, and particularly since the oil boom, there is need to analyze the economic, political, and social (ethno-cultural) structures inherited from its colonial past. To examine only economic changes, without considering its cultural history and conflicts in a multiethnic setting, the social demands that grew from that history, the differences in education, values, and access to power of the major political and ethnic groups, and the legacy of economic domination and dependence, would lead to a distorted understanding of the challenges to Nigeria's goal of autonomous development. Chapters 5–8 pay close attention to the economic and sociopolitical structures that have shaped Nigeria in the colonial and post-colonial period. In so doing, we shall be in a better position to fully understand the types of changes that have taken place in Nigeria since the oil boom of the 1970s, and also be able to provide the explanations for their occurrence.

NIGERIA TODAY: A PRISONER OF HER PAST

Nigeria abounds in a colorful variety of landscapes, vegetation, animal species, and human beings. It is situated on the west coast of Africa, on the Gulf of Guinea (which includes the Bights of Benin and Biafra), and extends north to the southern edge of the Sahara Desert. Nigeria has a south-north length of 650 miles and a west-east span averaging 700 miles, enclosing an area of 356,669 square miles (more than three times the size of the United Kingdom). It lies wholly within the tropics. It is bordered by the Republic of Benin (formerly the Republic of Dahomey) on the west, the Republic of Niger on the north, and the Republic of Cameroons on the east. Its population of 88 million means that one of every four Africans is a Nigerian (*Africa Magazine* 1981:80).

Once referred to as the "giant of Africa," Nigeria came on hard times early in the 1980s. Under the heading "Drowning in Unsold Oil: The Dangerous Collapse of Nigeria's Petroleum Prosperity," *Time* (1982:62) stated:

> Now that demand for petroleum is slumping everywhere, . . . the price of crude has dropped from $40 per bbl. to $28 per bbl. on the unregulated spot market. As a result, the economic outlook for the 90 million inhabitants of Nigeria, black Africa's wealthiest and most populous nation, has suddenly turned bleak.

According to the same article, Nigeria depends on oil exports for 90 percent of its foreign exchange earnings and 85 percent of its government revenues. Weakening worldwide demand for crude, however, forced the country to reduce production from 2.1 million barrels per day (b/d) to fewer than 0.9 million b/d during April 1982. As a result, estimated earnings from oil exports plunged from $1.35 billion per month to as little as $0.7 billion. To compound her problems, Nigeria was reported in the same article to be paying more than $1.8 billion each month for its imports, even though it had reserves of only $3 billion—less than enough to cover two months' import bills at the time.

The above report provides an indication of the economic problem Nigeria faces. It also demonstrates her total reliance on trade and capital attraction in her quest for national development. But how did Nigeria get into the situation described above? For an answer, not only do we have to examine her economic, social, and political history since attaining political independence in 1960, but we need to go back further, to examine her experience under colonial rule. Also, we cannot reduce this study to a mere economic analysis while ignoring the sociopolitical aspects of Nigeria's history. Her experience has shown that the economic development of a country is intertwined with her sociopolitical experiences. We cannot understand one aspect without examining the others.

This chapter, therefore, covers the period from 1860 to 1960. The objective is to show what types of political, social and economic structure the British colonial government developed for Nigeria during the colonial period. An understanding of these structures helps us to comprehend Nigerian experience with the oil wealth of the 1970s. Chapter 6 covers the first ten years after independence, and examines whether Nigeria achieved political and economic autonomy after independence or whether, despite political independence, the economy remained colonial in nature. Chapter 7 covers the period of the oil boom to 1984, and examines whether Nigeria's reliance on trade and capital accumulation as engines of growth has helped her to take off into sustained growth. It should be mentioned, however, that the history of Nigeria from the colonial period to the present is long and involved; only a broad outline can be provided in this book.

THE DRIVE TO INDEPENDENCE

The geographical area named Nigeria originally encompassed 250 ethnic groups with differences in their organization of political and economic life, in their cultures, and in their religious beliefs. James O'Connell (1967:132) states:

> No other colonial territory grouped people so large and diverse. The Hausa-Fulani and Kanuri and other peoples of the savana country, whose religious traditions were Muslim and who possessed developed administrative structures, were linked with tropical forest peoples like the Yoruba and the Edo, who possessed sacred chiefs and a hierarchical system of government, and Ibo, who were segmentary in their social organization and who were politically fragmented.

Despite this diversity in the composition of Nigeria, its traditional components show underlying similarities. Some of these are the extended family system, the concept of the High God, and the creation legends. These attest to the existence of a Nigerian traditional culture.

The ethnic groups of Nigeria had existed in the pre-colonial era as autonomous political entities. Some of them had achieved the status of empires, kingdoms, chiefdoms, or city-states, independent of European contact. There were the great kingdom of El-Kanem in Bornu, with a known history of more than 1,000 years; the Sokoto Caliphate, which for close to 100 years before the arrival of the British had ruled most of the savannah of northern Nigeria; the Hausa-Fulani emirates; the Ile-Ife and Bini kingdoms, whose art is recognized as among the most accomplished in the world; the Yoruba Empire of Oyo, which was once the most powerful state of the Guinea coast. Each was a sovereign state whose head had the power of life and death, conducted war and negotiated peace, and engaged in international diplomacy. As a result of their colonialism, in 1914 the British brought all these empires, kingdoms, chiefdoms, city-states, and the surrounding communities under the name of British Nigeria.

The first Europeans in Nigeria, however, were not the British but the Portuguese. They arrived in Lagos in 1472, and gave the city its name, which then was Lago di Kuramo (Niven 1967). They traded with the people, exchanging European products such as iron and metal objects for slaves and ivory. Such trade was mainly confined to the coast.

The British arrived later. After several futile expeditions by Europeans to find an easy route into the interior of Nigeria, Richard Lander and his brother John landed in 1831 at the town of Brass, under the sponsorship of the British government. Trading expeditions up the river Niger commenced. Even though the slave trade was declared illegal by Britain in 1807, the overseas trade actually ended only in 1850. Legitimate trading in ivory, gold, diamonds, palm oil, and palm kernels grew rapidly.

The British traders on the coast were in constant need of protection. They had sold arms to the local people, and these arms became a threat to them. They had to rely on the Royal Navy for support. The first direct British interference in Nigeria came in 1851, when a naval attack was directed against Lagos in order to force King Kosoko of Lagos off the throne for his failure to abandon the slave trade. He was subsequently replaced by King Akintoye, who was not only sympathetic to British interests but also dependent on Britain for his authority (Awolowo 1968; Crowder 1973). In 1861, Lagos was ceded to Britain and was administered as part of the Gold Coast (now Ghana). In 1862, Henry Stanhope Freeman was appointed first British governor for Lagos, which thus became the first section of Nigeria to come under British rule (Awolowo 1968).

By 1870, French traders could be seen on the Niger River, side by side with the British. In 1879, a trader named George (Goldie) Taubman brought order into the competing British commercial operations on the lower Niger by founding the United African Company, which absorbed most of the smaller companies trading in the region. He took the action in order to maximize profits for British interests and to present a common front to their French competitors (Awolowo 1968). The name of the British amalgamated firms was changed in 1881 to the National African Company Limited. In 1886, it was again changed to the Royal Niger Company Chartered and Limited. It was in this name that the firms obtained a charter from the British government on July 10, 1986. The charter in no way gave it the monopoly of trade on the Niger or in any other part of Nigeria, because Articles 26–29 of the Berlin Act of 1885 made provision for the free navigation of the Niger, its branches, its outlets, and its affluents for merchant ships of all nations carrying goods and passengers, and also provided for free coastal trading (Geary 1965).

However, the British charter gave the company power to administer by means of police, judges, and prisons, to make treaties, to levy customs duties, to raise taxes, to trade in all territories in the basin of the Niger and its affluents: in sum, to carry out all the functions of a government. In other words, the Royal Niger Company was both a trading and a governing concern (Awolowo 1968). The company also enjoyed military support in the form of river steamers with light guns and of an armed constabulary of 424 Hausa and Yorubas commanded by British officers (Geary 1965). Also in 1886, the British government proclaimed the Oil Rivers Protectorate over the Niger delta and established the Colony of Lagos.

In 1900, the British government took over the administration of the northern territories from the Royal Niger Company, revoked its charter, and declared the area the Protectorate of Northern Nigeria. At the same time, the Protectorate of Southern Nigeria was created to replace the Niger Coast Protectorate. Thus, from 1886, one year after the Berlin Conference on Africa ended, until 1900, and in order to avoid the cost of administering this part of its colony, the British government allowed the Royal Niger Company to rule the Niger and protect British interests. In 1906, the Colony of Lagos and Oil Rivers Protectorate became part of the new Protectorate of Southern Nigeria. In 1914, the Northern and Southern Protectorates were amalgamated to become Nigeria.

Between 1914 and 1922, there was the Nigerian Council for the Protectorate, which was mainly advisory. There was also the Legislative Council for the Lagos Colony. The two councils were abolished in 1922, and in their place a large Legislative Council was established; it included four elected members, three of them from Lagos and one from the port of Calabar. In 1923, the first election to the Legislative Council was held; the right to vote was based on an income and property qualification of £ 100 per annum (Geary 1965:271). The council legislated for the Colony of Lagos and the Western and Eastern Provinces of Nigeria, while the governor alone was responsible for legislating for Northern Nigeria (Awolowo 1968). In February 1924, the mandated territory of the Cameroons (a German colony before World War I) was joined to and administered with Nigeria. In 1939, the Northern and Southern Provinces were divided into the Northern, Eastern, and Western Provinces. It should be noted that until 1947, except for the abolished Nigerian Council, the people of the Northern Province did not participate in the Legislative Council (Coleman 1958). That no doubt accentuated the separate development of the North, and also shows the "divide and rule" strategy of the British during the crucial 1922–47 period. The British thus succeeded in ensuring that there was no common government that commanded the allegiance of all Nigerians. The division thus caused has continued until the 1980s.

The 1923 constitutional arrangements ended in August 1946, when the Richards Constitution was introduced. It was so called because the author, Sir Arthur Richards, then governor of Nigeria, handed the Constitution down to the people of Nigeria without any consultation. It provided for a central Legislative Council with power to legislate for the entire country, subject to the reserved powers of the governor. In addition, there was a House of Assembly for each of the three provinces. These assemblies were basically advisory in nature (Awolowo 1968). At first, the Constitution was planned to last for nine years, but political agitation by educated Nigerians led to its review only two years later.

The next constitution, the Macpherson Constitution, came into effect in June 1951. It was radically different from the Richards Constitution in that before its introduction, there was consultation with the people. It provided increased regional autonomy and gave Nigerians a greater share in policy formulation and in the direction of executive government action. It was this constitution that introduced representative government into Nigeria. It gave Nigerian leaders the opportunity to train in the art of modern government, and it prepared Nigerians for the day when Britain would transfer the administration of their country to them. As a result of the desire for greater regional government autonomy and the need for a more precise definition, division, and clarification of the functions of the central and regional governments, the first major constitutional crisis occurred in March 1953. In response to that crisis, a new constitution was introduced that established a federal system of government in 1954. It gave the regions greater autonomy in the Federation of Nigeria and made Lagos the federal capital (Coleman 1958). The new federation consisted of five parts: the Northern, Eastern, and Western Regions; the Federal Territory of Lagos; and the Territory of Southern Cameroons.

After another constitutional conference, at London in May– June 1957, self-government was granted to the Eastern and Western Regions in August 1957; the Northern Region became self-governing in 1959. During the negotiations for independence in 1959, Southern Cameroons decided to leave the Federation of Nigeria. Following further constitutional conferences in 1959 and 1960, the British government passed the Nigerian Independence Act, authorizing the queen in Council to draft a new constitution for an independent Nigeria. On October 1, 1960, the Federation of Nigeria became an independent and sovereign nation within the British Commonwealth. In the same year, it became a member of the United Nations.

In concluding this discussion of the political transformation of Nigeria from colonial status to independence, several points need to be borne in mind. First, the pre-constitutional and constitutional periods differ in the degree to which Africans could formally participate in their own governance, and the essential purpose of governance did not vary. Second, that purpose was to preserve and enhance a colonial relationship that served the economic interests primarily of trading companies and the British Empire. Third, the constitutional period evolved in response to British efforts to maintain the colony in the face of growing opposition from Nigerians. Fourth, independence was granted when British economic interests were no longer substantial, as a result of declining demand for Nigerian raw materials, cheaper substitutes, and growing Nigerian nationalism. The remaining sections of this chapter will expand on these points.

THE SOCIOPOLITICAL LEGACY OF COLONIALISM

The partitioning of West Africa among the European powers was not an isolated case. It was part of a wider movement by which nearly the entire continent of Africa was placed under European administration. By 1900, the whole continent, with the exception of Ethiopia, Liberia, and Morocco, was under various European administrations. The colonial powers included Belgium, Britain, France, Germany, Italy, Portugal, and Spain (Hatch 1970). They had divided the land and the peoples according to their commercial ambitions and their relative strengths.

The geographical boundaries drawn by the Europeans had little to do with the social, economic, and political realities of African life, and often disrupted established communities. The frontiers were frequently established by maneuvers of European armies. The Northern and Southern Protectorates of Nigeria and the Colony of Lagos were examples of what was transpiring throughout Africa at that time. Their amalgamation into one country called Nigeria in 1914 revealed the British view of Africans at that time. The British tended to regard all Nigerian societies as homogeneous. The assumption was that the Yoruba urban kingdoms, the Fulani-governed emirates, communities such as the Tiv, the village units of the Ibos, commercial ports of the Niger delta, and the mixed society of Lagos could all be "melted in a single pot." That assumption showed a total lack of understanding of the complex nature of the Nigerian societies. Colonial rule was thus imposed without enough understanding of the implications and consequences. As Coleman (1958:45) rightly points out:

> The artificiality of Nigeria's boundaries and the sharp culture differences among its peoples point up the fact that Nigeria is [a] British creation and the concept of a Nigerian nation is the result of the British presence.

The assumption that the region could be governed as a single nation served to produce deeper dissension among the various parts of the country.

At the start of the 20th century, Nigeria was divided into three administrative units—Northern and Southern Protectorates, and Lagos—with major difference among the areas and in their administration. Such variance and division prevented a united front against British rule. In

Lagos and Yorubaland, there was a continuation of a measure of social interaction between the British and the African elite. But such was not the case in the Southern Protectorate, where British rule was characterized by military force. As a result, the administration was resented not only by Nigerians in the region who had been accustomed to sharing in decision making, but also by the Nigerian elite in Lagos, who suspected an authoritarian regime was being home-grown in neighboring territories.

In the North, between 1900 and 1906, British Commissioner Frederick Lugard faced the basic problems common to most newly colonized areas, among which were lack of trained personnel and inadequate revenues. In addition, he had little trade and no cash production, and no ports to provide him with sources of taxation. He therefore relied on indigenous rulers and indirect rule (Burns 1969; Crowder 1973). By 1900, the Fulani emirates had been in existence for almost a century, and had become feudal oligarchies governing their Hausa-Fulani subjects. Each emir had a bureaucracy, based on adherence to Islam, that was responsible for tax collection, maintenance of law and order, and the administration of justice. Lugard thus had a well-established structure that he had only to convert or modify to conform to his objectives. His policy was to exercise power over all the emirs, coercing with military threat if necessary, and to use their system of administration and their officials as agents of Britain.

In the early days of British colonization, the railway was seen as a catalyst for economic development. Plans called for extending Nigeria's first railway, between Lagos and Ibadan, northward across the Niger River to be used by the Northern Protectorate. Hatch (1970:185) gives an account of how the divisive colonial administration policies worked to prevent consensus:

> Plans were actually accepted for the track's extension as far as Ilorin. But then the northern commissioner, successor to Lugard, proposed that the north have its own railway, from Kano to the Niger. This new railroad would enable northerners to import and export produce by rail and river, bypassing Lagos, which would consequently lose the anticipated customs revenues. To add insult to injury, because of the north's scanty revenues, a loan was raised for the project against the security of southern revenues.

Eventually, however, division became a financial liability to colonization. Despite the subsidy from the Southern Protectorate, the imperial government had to bear the major burden of providing a grant-in-aid, which rose from $201,400 in 1900 to $972,000 four years later (Hatch 1970). Therefore, the British administration sought to create a united Nigerian administration in order to pool revenues and costs. The practical result of the amalgamation was to enable the large revenues of Southern Nigeria to be used for the development of the entire country, to end the financial difficulties of Northern Nigeria, and to eliminate the imperial grant-in-aid, which had been reduced to £ 100,000 by 1914 and finally ceased in 1918 (Geary 1965:250).

The British policy made good financial sense, but it did not in any way address the political problem raised by unification. Such an important policy was formulated without due consideration for which system of administration was to be applied in a unified Nigeria. Was the system of advising and supporting feudal rulers while using their officials to be extended from the North to the South? Or were the more capitalist considerations prevalent in the South to be applied to the North? What would be the effects of merging traditionalist societies with individualistic and competitive societies? These kinds of questions were not what guided British colonial policy.

When Lugard returned to Nigeria to amalgamate the Northern and Southern Protectorates, he preferred the purely administrative function of government in the North and condemned the Southern administration for its encouragement of economic life and commerce. His attack throws into doubt whether Lugard ever saw his mission as one of building a modernized, self-governing Nigerian state. His proposal called for the merging of only the departments of the Railways, Marine, and Customs—all designed in such a way as to make the Southern commercial enterprise pay for Northern infrastructure. Other departments, including Medicine, Post and Telegraphs, and the West African Frontier Force, were organized in separate Southern and Northern branches with a joint head.

Lugard also sought to train chiefs in the South whom he could convert into feudal autocrats, establishing a political structure much the same as in the North (Crowder 1973). He suggested that the region should be divided into provinces with district officers exercising judicial and executive powers, as was customary in the North. Furthermore, he proposed that Lagos should be separated from the Southern Protectorate in order to limit the influence of the Legislative Council in Lagos so that it would legislate only for the city. His proposed substitute was a Nigerian Council that would consist of prominent men nominated by the government. The Colonial Office accepted Lugard's proposals, and the Northern and Southern Protectorates were amalgamated into a country named Nigeria in 1914. Although formally amalgamated, the two societies were deliberately kept apart, resulting in the isolation of the North from the impact of Southern economic activity and educational progress.

THE ECONOMIC LEGACY OF COLONIALISM

Before and during the British administration of Nigeria, the economy of the country could be divided into two distinct but tenuously connected sections. There was the economy of the natives, who were mostly farmers and traders, and that of the import and export merchants, the latter being mostly foreigners. The ratio between the two stood at thousands to one, with the farmers in the majority (Niven 1967). Since farming is seasonal, some farmers also engaged in trade. Certainly, no one starved while others stuffed themselves and threw away the excess in the traditional African societies (Igbozurike 1976). Poverty may have existed in pre-colonial Nigeria, but not to the extent that we now know it, with wealth and poverty polarized at two opposite extremes.

International trade in Nigeria did not originate with the British. In West Africa, slave trade flourished, and as early as 1441, slaves were sold in Lisbon by Portuguese traders (Crowder 1973). The British Parliament abolished the slave trade on March 31, 1808 (Awolowo 1968). Thereafter, and particularly after 1850, palm oil became the dominant export commodity (Crowder 1973). With the coming of British rule in Nigeria, the export trade continued and was even strengthened. By 1900, palm products made up 82 percent of total exports (Coleman 1958). The colonial governments were expected to pay their administrative expenses and to keep grants-in-aid from Britain to a minimum. In order to achieve these aims, export revenues had to be increased. Hence, increased production of cash crops was encouraged, not only for capitalist profit but also to increase government revenues. It is not uncommon to find the interests of the natives subordinated to the interests of merchants, manufacturers, and European financiers. The central roles of the colonial government were to promote the extraction of raw materials and the importation of British products, and to protect foreign investments in the country. The economic policy of the colonial government was the same as the expansionist policy of the British Empire. As Hatch (1970:204) points out:

> It was no objective of colonial rule to undermine the foreign dependence of colonial economies or to replace it by independent, self-propelling economies. The only part of the African continent in which independence was tolerated was in South Africa, and there the process was accomplished only after World War II as a result of an alliance between British, U.S., and white South African industrial finance. Elsewhere, while economic activity was often stimulated, the purpose of encouragement was either to assist the economy of the imperial power, or to promote the activities of its companies, or to swell the treasuries of colonial administrations. The methods of economic stimulation therefore had to fit these purposes.

When Britain established control over Nigeria, it did so for economic reasons. Its imperial policy often was rationalized as being in the best interests of the colonized. For example, Sir Hugh Clifford, a former governor of Nigeria, argued:

In the case of primitive peoples, unadulterated native rule is not popular or desired by the bulk of the natives. It means the oppression of the weak by the strong, the tyranny of might, the abnegation of law, the perform- ance of various bloody rites, and perennial inter-tribal strife—in a word, all the things which are most abhorrent to the principles of democracy. (Clifford 1918:13)

Sir Alan Burns, a former British colonial official in Nigeria, stated with re- ference to Nigeria that wealth was less the motive than was philanthropy:

and what justification was there for British rule? Imperialism is regarded in some quarters with such suspicion that it may be difficult to convince the prejudiced mind of the true motives that prompted the conquest of this vast territory. National acquisitiveness and commercial interests no doubt played a part, but in the case of Nigeria it may safely be said that the British entered on their great trust with reluctance and considerable hesitation, and that philanthropy was not the least of the influences that led us to take up the burden. (Burns 1969:306)

Despite the above views, the British actions in Nigeria were almost single-minded in the economic aim of exploiting the country's rich store of raw materials to run British factories and then to supply a potentially large market for products of British factories. Economic growth in colonial Nigeria was based initially on cash crops for trade, with the major cash crops being palm oil, timber, cocoa, cotton, peanuts, and kola nuts. Later, as early as 1904, tin was mined around Jos (Geary 1965). In order to induce growth in the trade sectors, a reorganization and expansion of the coun- try's communication and transportation network had to take place. Rail- way and road building became part of the program of the colonial govern- ment. Railway lines from Lagos to Ibadan were completed in 1900, and reached Kano in 1911; Jos was later connected to the main line in order to transport tin, peanuts, and cotton; and a line was run from Enugu to Port Harcourt, and eventually to Maiduguri to the northeast. By 1939, a net- work of 2,178 miles of track covered Nigeria (Coleman 1958).

Harbors were developed in Lagos and Port Harcourt. In 1913, deep- water berths were opened in Lagos; in 1926, additional ones were opened in Apapa. When the Udi coalfields near Enugu were discovered, the Port Harcourt harbor was constructed to serve the coal industry and empty the Benue Basin of its raw materials.

In 1900, roads were nonexistent in colonial Nigeria. By 1906, only 30 miles had been constructed. But after World War I, it was discovered that light Ford cars could travel on the rough African bush paths. As a result, by 1923, there were over 600 Ford cars in Lagos alone. These cars were, therefore, the impetus for road construction; by 1926, more than 6,000 miles had been built. By 1950, there were more than 28,000 miles of roads

(Coleman 1958). By 1959, the number of miles of roads had increased to 65,700 (Ukpong 1979). The railroads had been used for transporting tin, peanuts, and cotton. Road networks were therefore constructed as feeders of the railroad lines, so as to open up fresh commercial areas, such as that for cocoa. A closer look at the networks should reveal a crucial underlying factor: the transport lines all connected areas of raw material production to ports, which facilitated the export of raw materials.

From 1910 to 1957, coal was the mineral produced in greatest quantity, but tin was the most valuable on the world market; Nigeria is also the chief world producer of columbite (Buchanan 1966). Until its charter was revoked in 1900, the right to the extraction of all minerals in Nigeria was held by the Royal Niger Company. At the revocation of its charter, the company was paid various compensations. Its mineral compensation was to last for 99 years, so even when it no longer engaged in direct mining, it drew half the mineral royalties over much of the Northern Provinces until that share was purchased by the government in 1949 for £ 1 million sterling (Buchanan 1966).

In 1912, the West African Currency Board was established to introduce a coinage system to replace such traditional currencies as cowries, iron bars, and manilas. The standard for the new currency policy was 100 percent sterling exchange; that is, the currency issue must always be backed by reserves. The notes carried a promise to pay the bearer, on demand, the amount stated thereon; the amount required for such payment was to be a charge on the moneys and securities in the hands of the board or, failing them, on the general revenues of Nigeria (Geary 1965). That meant that all imports were to be paid for by current exports. The implication of this policy was that monetary reserves could not be used to finance internal development; all the reserves were tied up in supporting the currency, which in turn was limited to the existing reserves. The supply of money circulating in the country was thereby kept low. As of 30 June 1920, the face value of notes circulating in Nigeria was £ 2.5 million (Geary 1965). To increase the supply of money, the country had to increase export earnings and decrease import expenditure. The policy attracted foreign investment because it meant that inflation was virtually zero. But it also resulted in the absence of significant improvements in the living standards of the people.

This is not to say that Nigeria failed to benefit at all from colonial investment. For instance, the railway increased the ease with which Nigerian export crops could be moved from the interior to Lagos. It also facilitated greater internal trade: kola nuts and palm oil were sent north, and cattle transported south. The resultant increase in revenue led to some colonial funding for education and certain other social facilities. But these benefits were modest, unevenly distributed, and far less than those Nigeria furnished Britain.

One colonial policy in Nigeria that contrasted sharply with what was happening in East and Central Africa, and several French territories was the ceiling (in some cases outright prohibition) placed on land conveyance to Europeans (except in a few coastal areas). For the Northern Province, that policy was embodied in the Land and Native Rights Ordinance No. 1 of 1916 and No. 18 of 1918. The ordinance provided that occupancy rights granted to nonnatives should not exceed 1,200 acres if for agricultural purposes nor more than 12,500 acres if for grazing purposes. For the Southern Province, the Native Lands Acquisition Ordinance of 1917 proscribed nonnative land acquisition from a native except by approval, in writing, of the governor (Geary 1965). Despite pressures from European companies seeking to purchase land for plantations, even as early as 1907, the British Colonial Office refused to allow a plantation economy in Nigeria. The argument for a plantation economy had been that production could be increased by utilizing mass agriculture methods like those tried in the Congo Basin by Lever Brothers. Another argument was that the natives would be happier and more prosperous when their labor was directed and organized by Europeans.

Those were the arguments expressed in 1924 by Lord Leverhulme, chairman of Lever Brothers, to the governor of Nigeria, Sir Hugh Clifford, at a dinner held by the Liverpool Chamber of Commerce (Crowder 1973). But the experience in the Congo, where Lever Brothers had a large plantation, was one of brutal depredations that finally forced the Belgian government to take total control of the country just before World War I. Plantation economies in Africa had always meant forced labor, and usually resulted in resentment and hostility to the colonial government. Since the British were not willing to finance the necessary military defense of a plantation policy, the expansion of agricultural output was left entirely to Nigerians.

In the years between the two world wars, the economy remained highly dependent upon Britain and Europe. Fluctuations in export prices resulting from changes in European tastes and the terms of trade substantially reduced national revenues, and the recession in Britain halted internal development in Nigeria. In the first year of colonial government (1900), total colonial export earnings were $4.4 million. By 1921, the figure had increased to $21.8 million, then almost doubled to $39.7 million by 1926 (Hatch 1970:208). As a result of wartime shortages, those years were a raw materials boom period. But then demand for raw materials inevitably and suddenly halted. The effect was a drop in prices. Government revenues fell from $19.2 million in 1925 to $14.4 million in 1939. During the period, however, the Nigerian population grew from 20 million to 25 million. As a result, with the exception of the railway and a few other projects, the benefits of colonial development for Nigerians all but vanished.

During that period of declining trade, low revenues, and reduced prices, consolidations took place among trading companies in order to increase profits. In 1929, the United African Company (UAC) was formed by a merger of the Niger Company (which had continued its business after its charter was revoked in 1900, and was bought in 1920 by Lever Brothers) and the African and Eastern Trade Corporation, the product of another set of mergers. Because UAC was the largest of all the trading companies, it was able to fix prices it paid to producers and those paid to it for produce. It was even possible for it to fix prices and divide the market with large firms such as Compagnie Francaise de l'Afrique Occidentale and the Société Commercial de l'Ouest Africain. Their near monopoly position meant that they could ignore Nigerian needs and avoid the expenses of modernization. Their profits were mostly invested in European countries.

Formally, Britain declared a commitment to the principles of free trade in its relations with Nigeria. But these principles were usually discarded when British interests were in jeopardy. For instance, after World War I, discriminatory tariffs were placed on palm oil exported to Germany, in order to ensure that British processing plants received their required supplies. During the depression of the 1930s, the Japanese had begun to sell cheap cloth and other manufactured products to African colonies. That competition was detrimental to the Lancashire textile industry, which had already closed mills and laid off thousands. So, in 1934, the British sought to protect their textile firms and economy by imposing discriminatory duties on Japanese imports into each of the four British West African colonies. The British policy prevented the inhabitants of those colonies from enjoying cheaper clothing and other manufactured products.

Between World War I and World War II, trade, commerce, and economic enterprises, in addition to education and health services, were left to the care of private entrepreneurs and missionary agencies (Awolowo 1968). Not until after World War II was a development plan for Nigeria considered by the colonial government. The first ten-year plan for development and welfare was prepared by the colonial administration in Nigeria, in response to the Colonial Office's request that the colonies prepare development plans that would assist it in disbursing the Colonial Development and Welfare Fund, as proposed in the 1940 Colonial Development and Welfare Act. Under the act, the British government agreed to contribute £23 million to the total cost of the scheme, which was estimated at £55 million. The British contribution was to come from a £200 million fund earmarked for social and economic advancement in the colonies. The remainder of the funds was to come from external and internal loans raised by the colonial government in Nigeria and from internal revenues.

Various projects were covered under the scheme, ranging from small community improvements to major projects in education, health, and research facilities. About 25 percent of capital expenditure was earmarked for the improvement of water supply and medical services, while only 6.4 percent of all capital investments went into industrial and agriculturhal development (Tomori and Fajana 1979). Such emphasis on the building of social infrastructure was deemed necessary by the British government because it was felt that under the prevailing conditions in Nigeria, no balanced economic development plan could be successful unless the people were put in a position where they could participate in and take advantage of the economic activities (Okubadejo 1969). In order to raise the remainder of the funds required for the ten-year plan, the Development Loan Ordinance of 1946 was passed, authorizing the colonial governor of Nigeria to raise a loan of £8 million in England. In the same year, the Nigeria (Ten-Year Plan) Local Loan Ordinance was passed, empowering the governor to raise £1 million in Nigeria. These loans were changed to the general revenues and assets of Nigeria; and were to be repaid from Nigeria's revenues. Furthermore, the governor was authorized to establish a Loan Development Board that would make loans or grants, or both, to any authority or cooperative society approved by the governor in council for development projects.

Central planning was, however, very difficult because of the size of the country. The government decided to decentralize the planning process by first setting up regional development committees to advise the Loan Development Board on projects. That led to further decentralization of the Loan Development Board in 1949, when regional production development boards and regional loan boards were set up. The development boards were involved in improvement of roads, agricultural development, research, land settlement, and the establishment of industries (Niven 1967). Funding for them was provided by marketing boards that handled the sale of agricultural produce. Table 5.1 provides a breakdown of the amounts withdrawn by the marketing boards from the potential income of agricultural producers, from 1947–48 to 1961–62. One can see that producers of cash crops were the financing basis for development loans. These withdrawals, which ranged between 11.2 and 42.3 percent, turned out to be a disincentive for agricultural production in later decades.

Several factors led to the revision of the ten-year development plan of 1945 halfway through its life span. Most important, while the 1945 plan was highly centralized, the Nigerian government, in response to ethnic and other tensions, was not. The political division of the country into regions in 1947 particularly worked against strong central control. As the regions became more autonomous, deviation from the original plan widened. The government therefore introduced a revised plan for 1951-56.

TABLE 5.1. Government Withdrawals from Agricultural Producer Income, 1947–62 (£ million)

Product/Period	Potential Producer Income	Total Withdrawals	Withdrawals as Percent of Potential Income
Cocoa			
1947/48–1953/54	165.83	61.75	37.2
1954/55–1961/62	206.22	53.92	26.1
1947/48–1961/62	363.05	115.67	31.9
Peanuts			
1947/48–1953/54	98.78	39.55	40.0
1954/55–1960/61	149.66	22.34	14.9
1947/48–1960/61	248.44	61.90	24.9
Palm kernels			
1947–54	126.44	36.97	29.2
1955–61	116.56	31.34	26.9
1947–61	243.00	68.30	28.1
Palm oil			
1947–54	81.61	13.90	17.0
1955–61	72.42	18.54	25.6
1947–61	154.03	32.44	21.1
Cotton			
1949/50–1953/54	23.01	9.73	42.3
1954/55–1960/61	42.75	4.80	11.2
1949/50–1960/61	65.77	14.52	22.1

Source: Derived from Heillener 1970:123.

That plan was revised in 1954 when Nigeria became a federation and the regional governments took over those responsibilities that were within their jurisdictions.

During that period, the International Bank for Reconstruction and Development (IBRD) was invited to Nigeria to evaluate the resources available for future development, to study the possibilities for development of the major sectors of the economy, and to make recommendations for practical steps to be taken in the implementation of development plans. In its report the mission concluded that there were too many development boards, and that their functions were ill-defined and overlapping

(IBRD 1955). As a result, in 1955 the government created the Federal Loans Board, the Western Region Finance Corporation, and a development corporation in each region. It also created the National Economic Council (NEC), which served as a forum for discussing common economic problems of the regions.

When the original ten-year development scheme expired in 1955, the British government extended it for another five years, until March 1960 (Ekundare 1973). NEC did not have any input into the 1955–60 plan because it had been formulated prior to NEC's existence. In September 1958, a Joint Planning Committee was established to advise NEC. This was necessary because NEC members had other departmental and political responsibilities, and therefore could not meet as frequently as necessary (Tomori and Fajana 1979).

During the 1955–62 plan period, total planned public capital expenditure was £N-660.2 million. Of this amount, £N-378.4 million (57.3 percent) went into the economic sector, with transportation along receiving 38.7 percent. But, in contrast with the earlier Ten-Year Plan, in which water and health received 25 percent, under the 1955–62 plan they received only 9.8 percent of total capital expenditures. Administration was allocated 14.1 percent; the 1954 constitutional changes had given rise to the creation of regional bureaucracies and, thereby, the need for administrative infrastructure (Tomori and Fajana 1979). Table 5.2 contains the breakdown of the 1955–62 plan.

Development planning in Nigeria up to independence was an instrument used to foster British economic interests. The Ten-Year Plan came to life because the right to empire was being challenged. Within British government circles, there was concern about the administration of the colonies following the 1938 riots in the West Indies, which prompted the British government to set up a royal commission to investigate conditions there (Crowder 1973). The commission discovered that one of the main causes of the riots was the very backward condition of the islands. That prompted the British government to earmark a fund for the West Indies and then to pass the more general Colonial Development and Welfare Act for the other colonies. Hence, the Ten-Year Plan was not conceived by the British colonial government for the advancement of Nigeria, but was a by-product of the unrest in another part of the British colonial system. Under the Ten-Year Plan little provision was made for industrial development: that would have created competition against British exports to Nigeria. Also, in the agricultural sector, attention was concentrated on a few export products (Federation of Nigeria, 1961:6).

During the 1955–62 plan period, emphasis was on the transport sector in order to expand and link the areas of raw materials production, and to expand the distribution of imports. Therefore, the centerpiece of the

TABLE 5.2. Sectoral Distribution of Public Sector Capital Investment: 1955–62 Development Plan

Sector	Percent of Total
Economic	57.3
Agriculture, livestock, forestry, fishery	5.7
Trade and industry (including mining and quarrying)	3.2
Transport	38.7
Communication	4.4
Other	5.3
Social	24.0
Education	7.2
Health	4.2
Town and country planning (including sewerage, drainage, refuse disposal, housing)	0.8
Water (other than irrigation)	5.6
Other	6.2
General (including admin., defense, security)	14.1
Financial obligations	4.6

Source: Tomori and Fajana 1979:140.

colonial policy was the infrastructure designed to induce private investments, especially in agriculture and industry. It was similar to what Hirschmann (1958) calls a "permissive" sequence in an unbalanced growth strategy: that of concentrating investments in projects such as roads, communications, electricity, hospitals, and the like, in the belief that investments in directly productive activities will follow as a result of increased profitability.

PETROLEUM IN THE COLONIAL ECONOMY OF NIGERIA

As already discussed in this chapter, development planning under the colonial government in Nigeria focused on the expansion of international trade and the protection of foreign investment, and served as a means for exploiting Nigerian resources. As a result, during the colonial period, both agricultural and mineral resources were extracted and exported, especially to Britain. All along, the British colonial government had been interested in petroleum as well as the other exports discussed so far. In fact, in 1921, two oil exploration licenses were granted in the Southern Provinces. The first was to the D'Arcy Exploration Company for an area west of the Niger River to the Dahomey (now Republic of Benin) boundary, and for 50 miles inland from the sea. That license was allowed to lapse in February 1923. The second license was granted to the Whitehall Petroleum Company Limited, for an area that extended from the Akassa mouth of the Niger River, having the Niger as its western boundary, to the eastern boundary of the Cameroons, and from the sea in the south to seven degrees north latitude. The company allowed its license to expire after its geological survey.

No serious exploration for petroleum was undertaken until 1937, when the Shell-BP Petroleum Development Company of Nigeria Limited set up a camp at Owerri in Eastern Nigeria and began some preliminary investigations. World War II interrupted their work, and little progress was made until several years later (Ekundare 1973).

After World War II, Shell-BP resumed exploration for oil, but none was struck until November 1953. By 1960, prospecting for oil had been carried out over a number of years in the provinces of Ogoja, Onitsha, Owerri, Calabar, Rivers, Delta, Benin, and Ondo. Besides Shell-BP, the other petroleum companies operating in Nigeria by 1960 included Mobil Exploration Nigeria Incorporated; Tennessee Nigeria Incorporated, a subsidiary of Tennessee Gas; Gulf; Eastern; and Standard Oil of New Jersey.

Between 1951 and 1959, 70 exploration wells were completed, but only 36 produced any crude. The first of these producible wells was discovered in January 1956 at Oloibiri, 60 miles west of Port Harcourt. Ten months later, the second producible well was discovered at Afam, about 20 miles east of Port Harcourt. Since then, wells have been discovered at Bomu and Ebubu in Ogoni Division, and Ughelli in Delta Province. In 1957, two pipelines were laid: one from Oloibiri to Port Harcourt, and the other from Afam to Port Harcourt. The first shipment of petroleum oil, 9,000 tons, left Port Harcourt by tanker in February 1958 (Buchanan and Pugh 1966; Ekundare 1973).

Crude oil production in thousands of barrels per day amounted to 5.1, 11.2, and 17.4 during 1958, 1959, and 1960, respectively. Therefore, up to 1960, petroleum was not a major source of revenue for Nigeria, but was already showing that it was a growing sector.

THE GENERAL ECONOMY BY 1960

During the 1950s, Nigeria's GDP grew rapidly from £₦524 million in 1950 to £₦1.59 billion in 1955; by 1960, it had reached £₦2.4 billion (IMF 1984). Virtually all the sectors of the economy grew rapidly, agriculture being the key sector. Indeed, growth varied throughout the colonial period, depending on the condition of agricultural exports. As independence drew near, agriculture, after a century of colonial development, accounted for over 60 percent of national income.

By 1960, the major industries in Nigeria included food canning, brewing, foundry works, textile manufacturing, furniture manufacturing, palm oil processing, sawmilling, rubber processing, cement manufacturing, pottery manufacturing, plywood, cigarette manufacturing, soft drink bottling, crude petroleum extraction, boatbuilding, and metal doors and windows manufacturing. These were mostly foreign-owned, with some jointly owned by regional governments and foreign investors.

Table 5.3 shows the imports and exports of colonial Nigeria between 1921 and 1960. It is evident that a foundation was being laid for what was to come in the 1970s and the 1980s. In 1946, Nigeria was enjoying a favorable balance of payments, but by 1955, it started to import more than it was exporting. It should be borne in mind that most of the imports came from the United Kingdom.

TABLE 5.3. Total Exports and Imports, 1921–60 (£ million)

	Exports	*Imports*
1921	8.3	10.2
1925	17.2	13.9
1929	17.1	13.2
1935	11.6	7.8
1938	9.7	8.6
1946	23.7	19.8
1955	129.9	136.1
1960	165.6	215.9

Sources: Crowder 1973:263; Ekundare 1973:225–26; Geary 1965:256–57.

The major sources of funding in Nigeria up to 1960 were customs and excise, direct taxes (income tax and corporate tax), colonial development welfare grants, and internal and external loans. The most important source among these was customs and excise. In 1946, it accounted for about 43 percent of total revenues; it increased to 51 percent in 1950, to 71 percent in 1955, and to 73 percent in 1960.

For her economic development projects, Nigeria needed capital investment that exceeded her capacity to invest. During the period between the world wars, however, the government had been reluctant to utilize debt financing as a fiscal instrument for development. The construction of the railway was perhaps the most significant exception. Soon after 1946, however, both foreign and internal loans were sought. During the period from 1946 to 1960, external loans actually declined: from £ 24.9 million in 1946–47 to £ 20.5 million in 1960. In contrast, locally or internally raised debts increased from £ 0.3 million in 1946 to £ 13.6 million in 1960. External loans were floated on the London money market, while internal loans were from individual Nigerians, banks, local authorities, insurance companies, cooperative marketing and thrift societies, and other commercial companies. By 1960, interest payments amounted to £ 2.5 million, 3 percent of total federal revenues.

In 1913, there were 44 elementary schools—but 14,611 Moslem schools—in Northern Nigeria. In contrast, there were in the same year 534 schools—elementary, primary, and secondary—in Southern Nigeria. During the development program of 1955–60, free primary education was introduced in Lagos, Eastern Nigeria, and Western Nigeria. At the end of 1960, there were 112 primary schools in Lagos, attended by 74,468 pupils; 6,540 primary schools in Western Nigeria, for 1,125,000 children; 6,451 primary schools in Eastern Nigeria, with 1,431,000 children; and 2,600 primary schools in Northern Nigeria, for about 283,000 pupils. Furthermore, by 1960, Nigeria had four technical institutes: Yaba Technical Institute, Enugu Technical Institute, Kaduna Technical Institute, and College of Technology, Ibadan. In addition, there were a number of trade schools in major towns. Also, until 1960, students at University College Ibadan were awarded degrees by the University of London under a special arrangement. The other university in the country (established in 1960) was the University of Nigeria at Nsukka. Other institutions of higher learning included the Nigerian College of Arts, Science, and Technology, with branches at Enugu, Ibadan, and Zaria.

By independence in 1960, traces of the divisive nature of the colonial administration were still evident in the unbalanced educational development between the North and the South. The total number of schools in the North was then roughly one-third of the number of schools in the Western Province alone. It will therefore take several decades to rectify the results of the British "divide and conquer" strategy.

SUMMARY

From all the above, one can see that the drive to independence was a long and difficult one for Nigeria. The peoples of Nigeria, comprising several ethnic groups and languages, suddenly found themselves a country, not of their own free will but as a creation of British colonial power. Like other colonial countries, economic growth in colonial Nigeria was based on the expansion of cash crops, trade, and import-export transactions. Agriculture was the leading sector in Nigeria up to 1960. Petroleum had not yet become a major source of revenue, even though exploration was already in full swing. Also, several policies of the colonial government were formulated to attract foreign investment at the expense of the welfare of the indigenous peoples of Nigeria.

During the colonial era, a mixed economy was imposed on the country, and a policy of unbalanced growth was pursued. Emphasis was placed on developing overhead capital with the hope that other productive investments would follow. After independence, Nigeria continued to pursue national development based on strategies very similar to those of the British colonial administration. These strategies are examined in the next two chapters.

6 Period of Transition, 1960–70

INTRODUCTION

The 1960s were years of extraordinary change for the newly independent country of Nigeria. Neither the British government, which handed over power, nor the indigenous leaders who took the power from the British in 1960 anticipated the structural crisis that would accompany political independence. At independence, Nigeria was hailed as an example of multi-ethnic coexistence, of peoples joined together in a common political system. It became an example of the federal system of government and of democracy in the developing world. The events of the 1960s, however, threw the example into doubt and challenged the claim of a newly formed national cohesion. Before the decade was over, an attempt was made to break the Nigerian federation, which led to a bloody civil war.

By the end of the 1960s, the foundation had been laid for petroleum export to take over from agricultural exports as the chief source of government revenues and foreign exchange. It would also displace export agriculture as the foundation for planned development. The issue to be examined in this chapter is development, with political, social and economic structures taken as the key to development. To provide insight into the problems of post-colonial political and economic development, this chapter analyzes Nigerian political changes during the period 1960–70, and then examines the socioeconomic developments and the role of petroleum revenues during the same period.

POST-COLONIAL POLITICAL TRANSITION

Beginning in the 1930s, nationalist leaders such as Chief Obafemi Awolowo, Dr. Nnamdi Azikiwe, Herbert Macaulay, and Ernest Ikoli spearheaded the Nigerian drive for self-government (Coleman 1958). To achieve their goal, each leader engaged in political organizing. The first political party to be organized was the Nigerian National Democratic Party (NNDP), with Herbert Macaulay as the founder[1] (Eleazu 1977). Even though the organizers were nationalists, the party functioned only in Lagos. During the 1930s, another political party, the Nigerian Youth Movement (NYM), was formed. It was believed to be more representative of Nigeria. Its leaders included Dr. Abayomi, Ernest Ikoli, H. O. Davies, Nnamdi Azikiwe, Obafemi Awolowo, S. L. Akintola, S. O. Sonibare, and J. Martins. By 1938, it had replaced the NNDP. However, after World War II, the NYM was in disarray. There are conflicting accounts of what led to the demise of the party. Some believed that the Yoruba leaders who had dominated the party were resentful of Nnamdi Azikiwe's rise to the top of the party. But there were others who believed that Azikiwe was a "bad follower" who broke with the party when his leadership was denied (Eleazu 1977).

During the rift in the NYM, the National Council of Nigeria and Cameroons (later renamed National Congress of Nigerian Citizens; NCDC) was founded by Azikiwe. Herbert Macaulay became the new party's first president and Azikiwe was secretary general. On Macaulay's death in 1946, Azikiwe became the party's president. Although a nationalist party, it was unable to recruit followers from the other ethnic areas or to extend its influence to the other areas. Consequently, between March 1950 and March 1951, the Yorubas organized their own party (Action Group) under Chief Obafemi Awolowo (announced March 21, 1951) (Coleman 1958). In 1949, the Hausas established the Northern Peoples' Congress (NPC), with Alhaji Abubakar Tafawa Balewa as its main originator. The NPC, which at first was a cultural group, was dominated by traditionalist rulers (Coleman 1958). In 1950, Alhaji Aminu Kano led a faction of the NPC to form the Northern Elements Progressive Union (NEPU) (Coleman 1958). Soon after, the NPC was reorganized as a political party in which Alhaji Ahmadu Bello, the Sardauna of Sokoto, the Northern Region premier, became the dominant figure, with the support of the emirs. These were the main political parties as of October 1, 1960, when Nigeria attained political independence.

The first election for a federal and independent Nigeria was held in 1959. This government was to take over power from the British colonial government. Because no party received the majority, a coalition government was formed. It was made up of the NPC (134 seats) and the NCNC (89 seats), despite the fact that the NCNC had participated in an electoral pact with NEPU, an opponent of the NPC, in the Northern federal election. The Action Group, with 73 seats, was the opposition party in the central government (Eleazu 1977). Sir Tafawa Balewa, a Hausa of the Jere community near Bauchi, became the first prime minister of Nigeria. Nnamdi Azikiwe, president of the Senate, was sworn in on Novermber 16, 1960, as the first Nigerian governor-general and commander in chief of the federation, thereby succeeding the former British governor-general of Nigeria, Sir James Robertson (Crowder 1973).

Even though the NPC controlled the organs of the central government and the Northern Region, the NCNC ruled in the Eastern Region and the Action Group ruled in the Western Region. It was soon realized on all sides that it would be impossible to formulate national policy to speed up development and pool resources in the interests of long-term planning and development. The result was that each region demanded greater autonomy to shift power away from the center.

During the first five years of independence, those educated Nigerians who had gone to universities before taking jobs in the government and in the other professions, and who had acquired a national outlook, were disappointed and frustrated by the inefficiencies and ineffectiveness of administration at the center and the high levels to which corruption and nepotism had reached.

In 1965, the coalition of the NPC and the NCNC started to unravel. Important political figures belonging to the NCNC withdrew their support from the government and gave it to the opposition party, the Action Group, to form the United Progressive Grand Alliance (UPGA). A new coalition of forces from the East and the West was thereby formed against the powerful North. This development quickly led to a polarization of political forces in Nigeria. The UPGA was supported by a majority of Ibos and Yorubas, while the NPC continued to command the allegiance of the Hausa-Fulanis and was able to attract the support of important sections of the Yoruba party headed by Chief Samuel Akintola, who wished to share power with the North. Akintola had formed the Nigerian National Democratic Party (NNDP) in the previous year (Crowder 1973), and he helped forge a coalition with the NPC for the December 1964 general election under the umbrella of the Nigerian National Alliance (NNA), headed by Alhaji Ahmadu Bello (Niven 1967).

The 1965 election in the Western Region resulted in chaos amid claims of fraud. The people of the region refused to accept the results. The Western Region government, led by Akintola, announced that the NNA, had won 71 seats, and that the UPGA, led by Alhaji Adegbenro, had won only 17 seats. UPGA, however, claimed that it had won 68 seats and that the elections had been rigged. Adegbenro then decided to form his own government, and he was arrested (Cervenka 1971). There followed a massive destruction of property and loss of lives. In January 1966, with Prime Minister Balewa unable or unwilling to restore order in the Western Region, military intervention was virtually assured.

The important political features of Nigeria during 1960–65 were the attainment of political independence in 1960 and the failure of the existing political parties to form a viable alliance, despite claiming to represent the interests of the major ethnic groups. The political history of Nigeria serves as an illustration of a federation with vast historical, social, and cultural differences among its various elements. It also demonstrates the effects on a federation where the various elements professing to be bound by common interests are actually seeking the greatest share of the "national cake" for themselves, and where there is an inequitable sharing of power between the center and the regions. The military intervention of January 15, 1966, culminating in the civil war, was the result of the inter-ethnic tensions suppressed in colonial Nigeria.

THE CIVIL WAR AND THE RISE OF THE MILITARY

Before discussing the rise of the military in Nigeria during the 1960s, it is worth noting that on the international stage of sub-Saharan Africa, the Nigerian army was much appreciated and envied.[2] In December 1960, under the auspices of the United Nations, Nigerian troops were sent to strengthen the foundation of the newly independent Republic of Congo Leopoldville (Zaire) (Crowder 1973). In 1964, the Nigerian army was brought into Tanganyika (now Tanzania) to stabilize President Julius Nyerere's political power.[3] The presence of the Nigerian army in Tanganyika brought African nationalist ideology and international relationships to a new level of awareness. These events helped the reputation of the army among Nigerians as well.

By January 1966, the stage had been set for a revolutionary change in Nigeria. The political turmoil resulting from election rigging and other polling irregularities in the Western Region had reached a critical level. Therefore, on January 15, 1966, Maj. Chukwuma K. Nzeogwu and a small detachment of middle level and junior officers overthrew the civilian regime. Prime Minister Balewa, Premier Alhaji Ahmadu Bello of the Northern Region, Premier Samuel Akintola of the Western Region, and a number of military officers, including Brig. Samuel Ademulegun and Brig. Mai-Malari, were killed.[4] Nzeogwu was quoted as saying that they "wanted to get rid of rotten and corrupt ministers, political parties, trade unions and the whole clumsy apparatus of the federal system" (Eleazu 1977:6). In his view and that of his followers, these were the institutions that were impeding Nigeria's development.

Gen. Johnson Aguiyi Ironsi, an Ibo, who assumed the military command of the Nigerian armed forces in 1965, was not a party to the coup but was able to assume control of the government after the acting president of the federation, Dr. Nwafor Orizu, handed over power to the armed forces (Cervenka 1971). On January 17, 1966, Nzeogwu surrendered to Ironsi. The latter then proceeded to appoint a military governor for each region. None of the majors who masterminded the coup assumed power, but military rule was a direct result of their action.

Following the military coup, civilian political disorder vanished. However, a few weeks after Ironsi assumed power, the coup started to acquire an unexpected dimension in the eyes of many Nigerians, particularly non-Ibo ethnic groups, who began to view it as an Ibo conspiracy against all other ethnic groups. The majority of the officers who carried out the first coup were Ibos; all of the important military and political leaders who had been killed were from the Yoruba, Hausa-Fulani, and Bendel areas. No Ibo leader was killed during the first coup. Giving added weight to the other ethnic groups' suspicions was the fact that Ironsi was an Ibo. Also, it was alleged that 19 out of 23 army officers promoted during the period were Ibos (FRN 1968). Ironsi's plan, drawn up with a view to establishing a unitary state (partly in order to prevent further outbursts of violence), was strongly opposed. The situation was further complicated by the belief of the other ethnic groups that Ironsi tended to trust his Ibo counselors and relied on them to the exclusion of other groups.

By May 1966, the general political situation in the country was still deteriorating. Riots occurred in a number of towns in the Northern Region and many Ibos lost their lives. The armed forces were unable to control the chaos because they too had begun to divide along ethnic lines, and seemed to have given up their discipline of unity.[5] On July 28, 1966, at about 11 p.m., an army mutiny took place at Abeokuta, Western Nigeria, during which Major Obienu, Lieutenant Orok, and Lieutenant Colonel Okonweze, all from the Eastern Region, were killed (Cervenka 1971). By 9 a.m. on July 29, while Ironsi was visiting the military governor of the Western Region, Lt. Col. Adekunle Fajuyi, and before he was aware of the mutiny, another military coup took place in which he was assassinated together with his host (Crowder 1973). Gen. Yakubu Gowon (then a lieutenant colonel), a Christian from the north who belonged to none of the main ethnic groups, became the new head of state. He immediately reversed the decision to establish a unitary state and restored the federal system of government (Burns 1969).

The objective of the first coup supposedly was to remove the NPC from power and to end the corrupt practices into which the country had sunk under the leadership of civilian politicians. The second coup reflected the destructive effect of ethnic differences on the entire structure of Nigerian life. For the first time, Nigerians began to wonder about the genuineness of "nationalist ideology" that was supposed to be exemplified by the military. The July 1966 coup did not restore peace to the federation. Instead, as a result of the riots in some Northern cities, particularly in Kano, the massive emigration of the Ibos and other non-Northern people living in the North continued long after the second coup. Furthermore, the tensions prevailing between the East and the rest of Nigeria became more and more intense. It was becoming clear that a civil war was imminent. Several reconciliatory steps were taken, most notably the meeting of the Nigerian Supreme Military Council held at Aburi, Ghana, on January 4–5, 1967, to resolve the question of leadership and the principle of collective rule on the basis of the concurrence of all military governors in decisions affecting their regions.

On February 25, 1967, Lt. Col. Odumegwu Ojukwu, military governor of the Eastern Region, threatened to implement the Aburi agreement on his own if the federal government failed to implement it fully by March 31. But, rather than secede on March 31, Ojukwu promulgated Revenue Collection Edict 11 of 1967, which stated that from April 1, 1967, all revenues collected in the Eastern Region for or on behalf of the federal government would be paid to the treasury of the Eastern Region government. On May 27, Gowon announced the abolition of the four regions in favor of 12 states. On May 30, Ojukwu led the Ibos in the East to declare an autonomous and separate nation-state of Biafra.[6] The declaration precipitated a bloody civil war. The federal military government viewed the action in the former Eastern Region as an act of rebellion, and therefore ordered an economic blockade of the area, followed by a police action. War was declared on July 6, after Ojukwu's forces occupied the former Midwest and advanced on the Western Region. The war continued until January 12, 1970, when the Biafran forces surrendered to the federal forces. By 1970, when the civil war ended, Nigeria had a military force of 250,000, that is 25 times its size in 1960 (Kirk-Greene and Rimmer 1981).

The Nigerian civil war attracted several foreign powers and other interventionists. For the most part, the main attraction was Nigeria's rich oil reserves. Each side in the civil war believed that it could manipulate outside forces seeking to exploit these reserves. Because a large proportion of the oil reserves was in the East, Ojukwu sought to use oil as a political inducement in his attempt to create an independent Biafra. This meant that the British Petroleum Company, which then owned most of the exploitation rights of oil resources in eastern Nigeria, had to decide which side to support. The British (Labour) government, after a period of neutrality, decided to support the federal government. The French (Gaullist) government revived inter-European colonial competition by deciding to support Ojukwu. The Soviet Union seized the opportunity to enhance her friendship with Nigeria by supporting the federal government. The Portuguese took the side of Biafra. The United States formally remained neutral, but was believed to support the "One Nigeria" concept. African countries were divided along predictable lines based on whether they were afraid of potential Biafras imbedded in their own political structures. Countries such as Zambia and Tanzania, which supported Biafra, did so because they were not faced with problems of multi-ethnic coexistence involving competition for political power and control of economic resources.

Despite the outcome of the civil war, the original tensions within Nigeria have in no fundamental sense disappeared. Military solutions are not viable substitutes for economic and sociopolitical solutions.

POLITICAL INDEPENDENCE AND ECONOMIC DEPENDENCE: THE COLONIAL ECONOMY OF POST-COLONIAL NIGERIA

Before independence, the Nigerian economy was dominated by the trends and fluctuations of its major export goods. Agricultural exports were initially the engine of growth in the Nigerian economy. Between 1940 and 1960, changes in the agricultural sector occurred mainly as a result of farmers' responses to income incentives generated by the integration of the traditional agricultural economy into the world market. The increase in output in the sector came from the employment of surplus land and labor, and from the substitution of higher value export crops for food crops, without significant reorganization of the society or the introduction of new poduction techniques. After independence, the circumstances persisted.

Table 6.1 presents the GDP by sectoral origin for 1950–60 and 1960–70, and the average annual sectoral growth rates for 1960– 70. From 1960 to 1970, no growth took place in the agricultural sector. In fact, it declined at an average annual rate of 0.4 percent. For 1950–60, agriculture as a share of GDP was 64.3 percent, but during 1960–70, its share of GDP declined to 56.7 percent. After 1967, the sector was affected by the civil war. While agriculture was experiencing stagnation during the 1960s, the manufacturing sector expanded rapidly, at an average annual growth rate of 9.1 percent. Increased demands generated by this rapid growth of industrial production, as well as export agriculture, led to very substantial growth in the public utilities and construction sectors.

During the 1960s, the mining sector began to play the role of the potential leading growth sector in the economy. During that decade, it grew at an average annual growth rate of 20 percent. The mining sector (mainly petroleum) thus became the dominant development sector between 1960 and 1970, as oil production in Nigeria began on a significant scale. Apart from enjoying a freight advantage over the oil from the Persian Gulf because of Nigeria's closer proximity to world oil markets, the quality of

TABLE 6.1. Average Annual Growth Rate of GDP, 1960–70
(percent)

Sector	GDP By Sectoral Origin		Avg. Annual Growth Rate 1960–70
	1950–60	1960–70	
Agriculture	64.3	56.7	-0.4
Mining, incl. petroleum	1.2	3.5	20.0
Manufacturing	3.5	6.5	9.1
Construction	3.8	5.0	6.0
Electricity, gas, water	0.7	0.5	10.3
Transport/communications	5.5	4.8	-0.3
Trade, finance	14.3	12.7	—
Public admin./defense	3.3	3.9	13.6
Others	3.5	6.3	1.4
Total	100.0	100.0	3.1

Note: Figures are rounded.

Source: World Bank 1983b: I, 134–35.

Nigerian crude is well suited to refineries in the United States and western Europe. It is light and nearly sulfur-free; hence, demand was high.

Exports of oil began in February 1958, following completion of the pipelines to Port Harcourt in 1957 (Buchanan and Pugh 1966; Ekundare 1973). By 1960, Shell-BP was producing some 17,500 barrels per day (b/d). A tanker terminal and related facilities were completed at Bonny in 1961, allowing production to rise to over 46,000 b/d. Nigeria's first refinery was established in 1965 at Alese Eleme, near Port Harcourt, as a joint venture of the Nigerian government (55 percent), Shell (22.5 percent), and British Petroleum (22.5 percent). It was operated by the Nigerian Petroleum Refinery Company.

Table 6.2 presents crude oil production in Nigeria between 1958 and 1970. With large-scale exploration in the early 1960s, reserves, production, and exports increased rapidly. The completion of the trans-Niger pipeline in 1965 allowed oil from fields in the Midwest to flow to Bonny. In the same year Gulf Oil began lifting crude from its first find, which also happened to be the first offshore field in Nigeria. Hence, crude petroleum

TABLE 6.2.   Crude Oil Production, 1958–70 (1,000 barrels)

Year	Production
1958	5.1
1959	11.2
1960	17.4
1961	46.0
1962	67.5
1963	76.5
1964	120.2
1965	274.2
1966	417.6
1967	319.1
1968	141.3
1969	540.3
1970	1,083.1

Source: OPEC 1977:217.

production rose to neary 600,000 b/d by mid-1967, prior to the Nigerian civil war, which caused all onshore areas to cease operation. Despite the war, oil output recovered rapidly from the 1968 low of 0.14 million b/d to reach an average of 1.1 million b/d in 1970. By 1970, natural gas was a by-product of crude oil extraction, and reserves of natural gas roughly followed the growth of recoverable crude oil reserves. The average amount of associated gas produced with crude oil was about 750 cubic feet of gas per barrel of crude. In 1970, about 98 percent of gas produced was flared.

Nigeria's economic development was, and remained, in the first decade of independence, highly dependent upon conditions in world markets. This dependence quickly created problems for post-colonial Nigeria. Table 6.3 shows Nigeria's exports and imports between 1950 and 1970. During the period, Nigeria's exports increased by 234 percent and imports increased by 326 percent. Exports of peanuts, cocoa, timber, cotton, and rubber increased rapidly, leading to a diversified agricultural export economy. However, the steady growth in export volume was not followed by corresponding growth in export value. It was rather erratic be-

TABLE 6.3. Exports and Imports, 1950–70
(million of naira)

Year	Exports	Imports	Ratio Exports/Imports
1950	266.7	219.3	1.22
1955	273.2	368.1	0.74
1960	278.5	481.0	0.58
1961	341.4	487.0	0.70
1962	366.8	456.4	0.80
1963	379.0	455.7	0.83
1964	423.3	547.0	0.77
1965	567.4	590.0	0.96
1966	585.5	572.2	1.02
1967	517.7	556.8	0.93
1968	462.4	517.9	0.89
1969	599.5	629.2	0.95
1970	890.5	934.7	0.95

Note: Ratio calculated by author.

Source: World Bank 1976:179.

cause of price fluctuations (IMF 1984:454–55). At the end of World War II, Nigeria enjoyed favorable terms of trade. After 1954, however, the favorable terms of trade suddenly vanished; prices of Nigeria's exports started to decline while import prices continued to rise (World Bank 1976:178–79). Imports expanded rapidly, from N 219.3 million in 1950 to N 934.7 million in 1970. Import-substitution industries in processed food, beverages, and textiles, and restrictive import policies during the civil war prevented further expansion of imports. Still, food and raw materials imports as a percentage of all imports increased from 19.9 percent in 1960 to 21.5 percent in 1970 (World Bank 1976:458). Also, the rapid growth of the industrial sector caused imports of machinery and equipment to increase from 24 percent in 1960 to 37.4 percent in 1970 (World Bank 1976:458).

These changes in the trade pattern produced considerable fluctuations in Nigeria's balance of payments, as shown in the ratio of exports to imports in Table 6.3. Trade deficits continued annually until 1966, when import substitution and expanding petroleum export brought a surplus. But the trade surplus did not give the true picture of foreign exchange remaining in Nigeria. Most of the import-substitution industries were financed to a large extent by foreign capital. Furthermore, petroleum production in Nigeria was controlled by foreign companies and, like Nigerian exports in general, the return to Nigerians was low.

TABLE 6.4.   Private Foreign Investment, 1960–68

Year	£ Million	Percent Invested in Oil
1960	24.0	—
1961	27.3	25.0
1962	17.7	42.0
1963	37.9	33.0
1964	63.0	57.0
1965	37.0	47.0
1966	34.9	83.0
1967	49.4	92.0
1968	60.8	71.0

Source: Dean 1972:208–09.

Table 6.4 shows the concentration of private foreign capital in the oil industry, particularly since 1961. In 1967, private foreign investment in oil accounted for 92 percent of all private foreign investment in Nigeria. Hence, investment income earned by non-Nigerians (factor payments abroad) increased from N 32.4 million in 1960 to N 635 million by 1970. Petroleum, therefore, remained a typical enclave industry whose contribution to the economy was limited largely to its contribution to government revenues and foreign exchange earnings. Crude petroleum exports rose from N 9 million in 1960 to N 510 million in 1970 (IMF 1984).

Table 6.5 reveals that while recurrent revenues increased by 37 percent, from £N 111.8 million in 1961 to £N 153.2 million in 1966, the recurrent expenditures of the federal government increased by 83 percent, from £N 51.7 million to £ N 94.4 million. Federal government surplus

TABLE 6.5. Federal and Regional Governments' Recurrent Revenues and Expenditures, 1961–66 (million £N)

	1961	1966	Average Annual Percent Change
Recurrent revenues			
Total	111.8	153.2	6.2
Retained by federal govt.	78.4	92.5	3.0
Transferred to regional govts.	33.4	60.7	13.6
Fed. govt. recurrent expenditures	51.7	94.4	13.8
General administration	10.8	21.7	16.8
Defense/internal security	12.9	23.2	13.3
Economic and social services	18.2	21.9	3.4
Debt charges	5.1	22.6	57.2
Other	4.7	5.0	1.1
Surplus/deficit	+26.7	-1.9	
Regional governments			
Recurrent expenditures	57.8	86.1	8.2
General administration	12.6	18.1	7.3
Economic and social services	37.5	58.4	9.3
Other	7.7	9.6	4.1
Surplus/deficit	-23.1	-12.9	
Fed. and regional govts.			
recurrent exp.	109.5	180.5	10.8
Public debt			
Internal	49.4	173.2	41.8
External	42.9	70.2	10.6
Total	92.4	243.4	27.2

Source: Tims 1974:136–37, 234.

dwindled from £N26.7 million in 1961 to a deficit of £N1.9 million in 1966. While federal expenditures on essential economic and social services rose at an annual rate of 3.4 percent, debt charges, general administration, and defense and internal security grew at annual rates of 57.2, 16.8, and 13.3 percent, respectively. The Nigerian civil war exacerbated an already difficult fiscal situation. Not only did it cause the defense operating expenditure to grow, but it also caused the stoppage of all planned economic development projects from 1967 to 1969. Defense capital and operating expenditure as a proportion of general government revenues rose from 4.8 percent in 1960 to 38.1 percent in 1970, while civilian consumption decreased from 71 percent to 50.9 percent during the same period (World Bank 1976:435). These cuts in civil consumption and substantial internal borrowing provided the funding for the civil war. As a result, internal outstanding debt increased from ₦514.2 million in 1968 to ₦1 billion in 1970, while foreign debt decreased from ₦179 million to ₦175.4 million during the same period (IMF 1984).

From the above, it is evident that economic growth in the 1960s was achieved without significant structural transformation of the economy.

THE FIRST POST-COLONIAL NATIONAL DEVELOPMENT PLAN

Despite the existence of political tension and fiscal difficulties, Nigeria was able to implement its First National Development Plan 1962–68 (hereafter referred to as the First Plan). The architect of this plan was the head of the Economic Planning Unit in the Federal Ministry of Economic Development, Dr. Wolfgang F. Stolper. He had come to Nigeria from the Economics Department of the University of Michigan in 1960, under the auspices of the Ford Foundation (Stolper 1966; 1970). Stolper accepted rapid growth in production per capita as the basic purpose of development planning (Stolper 1970). To him, the questions of how to achieve that growth were not only logically distinct but also factually separate from questions of what to do with it. According to him, wherever possible, decisions on resource utilization must satisfy the test of economic profitability. The limits on health care and public education, for instance, were to be set by the productiveness of the economy; the more the uses of resources for economic purposes satisfied the profitability test, the more resources would be left over for noneconomic purposes (Stolper 1970). In his view, the major problem confronting African businessmen and farmers was more one of productivity of labor and management than one of capital shortage (Stolper 1970).

According to the First Plan, Nigerians should have before them "the ultimate goal of economic independence" while being willing to accept assistance in manpower and materials so as to "advance the time when the country may become self-reliant," and able to proceed from the "take-off stage" into a sustained growth (FRN 1964:1–3). The plan projected a real growth in GDP of at least 4 percent per annum. Emphasis was placed on the productiveness of the economy and the autonomy of the country, whereas distribution of income was mentioned sparingly. As Stolper (1970:225) points out:

> a very good case can be made that premature preoccupation with equity problems will backfire and prevent any development from taking place. Furthermore, in no African country do we have really good information on existing distribution of income or wealth.

The First Plan called for expenditure of ₦1.351 billion. Out of that amount, 67.8 percent was earmarked for the economic sector, 24.4 percent was allocated to the social sector, and administration received 7.2 percent (Tomori and Fajana 1979:140). About 50 percent of the funding requirement for the plan was expected to be financed through foreign aid; during its implementation, however, only 25 percent was received (Dean 1972). Also, most foreign loans and grants had strings attached; they were tied to specific projects and therefore could not be used for other projects, regardless of their priorities in the plan. Furthermore, specific requirements of foreign aid donors had to be incorporated into the plan. These conditions caused difficulties in raising the external funding needed for the First Plan. As for the internal source of funding for the plan, the government had to rely on the revenue accruing from recurrent surpluses and the statutory corporations (including the marketing boards). Domestic borrowing provided a substantial part of the internal funding (FRN 1964; 1970).

Table 6.6 presents the distribution of public sector capital investment by sector during 1962-68. As under the colonial plan periods since 1946, infrastructure investment received high priority: transport, 21.3 percent; electricity, 15.1 percent; trade and industry, 13.4 percent; agriculture, 13.6 percent.

Among the projects planned was an iron and steel plant. At the end of the plan period, work had not progressed beyond the preliminary studies (Berger 1975; FRN 1970). The second major project planned under the First Plan was a petroleum refinery at Alese Eleme near Port Harcourt, which was completed in 1963. It had an annual production of 1.5 million tons, which was increased to 2.5 million tons per annum after the civil war (Berger 1975:88).

At the end of the plan period, less than one-third of the ₦180 million earmarked for industry had been spent, while another ₦10 million was invested in nonviable industries—a result of faulty project planning (Berger 1975:88). The low rate of performance was caused by lack of a concrete list of projects and of clearly defined measures for their promotion.[7]

In evaluating the success of the plan,[8] one ought to judge it on the basis of its aggregate nature; the focus was on the productiveness of the economy and the autonomy of the nation, even though distribution of in-

**TABLE 6.6.   First National Development Plan, 1962–68:
Sectoral Distribution of Public Sector Capital Investment**

Sector	Percent of Total
Agriculture	13.6
Trade and industry (incl. mining)	13.4
Electricity	15.1
Transport	21.3
Communication	4.4
Water (other than irrigation)	3.6
Education	10.3
Health	2.5
Town and country planning (incl. housing)	6.2
Cooperative and community devt.	0.6
Labor, social welfare, sports	0.7
Information	0.5
Judicial	0.1
General admin., defense, security	7.1
Financial obligations	0.6

Source: Federation of Nigeria 1961:41.

come among people and regions was mentioned in the objectives. On the basis of the productiveness of the economy and the rapid growth of GDP, the First Plan was right on target. It projected an average annual growth rate of 4 percent; the actual annual average growth rate of GDP between 1960/61 and 1965/66 (before the civil war) was 4.8 percent (Rimmer 1981:44; Tims 1974:211). On the national autonomy question, one only need reexamine the sectoral distribution of the plan. It is not difficult to see that heavy investment in the transport sector could not bring national autonomy. Chapter 5 has already shown how investments in the transport and other infrastructure sectors served to facilitate trade expansion, to induce, and, in effect, subsidize foreign capital. It is difficult, then, to believe that the motives for infrastructure investments changed during the drawing and implementation of the First Plan to fostering national autonomy. Also, trade and capital accumulation do not in themselves produce national autonomy.

The First Plan showed that even though Nigeria was independent, its national planning efforts continued to be colonial in nature. Like the pre-independence plans, it was drawn up by foreigners who were from capitalist nations where the economies are not planned, and who lacked the knowledge of the histories and local customs of the Nigerian people. The First Plan was, therefore, a direct transfer of a Western model and values. As to the question of distribution of income, there was no clear policy in the plan as to how that was to be carried out. It was stated merely to keep the masses happy. Investments in transport, electricity, water, and communication tend to subsidize the incomes of urban dwellers rather than of rural dwellers. In this respect, the redistribution could be said to be from the rural dwellers to the urban dwellers. Also, by emphasizing projects that were mainly complementary to rather than competitive with private foreign investment, the plan was designed to encourage foreign investment, particularly in manufacturing. Hence, the emphasis on building infrastructure (roads, bridges, railways, telecommunications, electricity, ports, and air transport), and relatively few manufacturing projects (Dean 1972). As is argued by Aboyade (1968), the leading politicians did not understand the implications of economic planning or the nature of economic development; their main concerns were with regional power struggles, and the power of political parties and individuals. Plans were secondary to the political system.

Apart from the above, the First Plan had been criticized for not bringing into use the unemployed productive capacity of the country (Schatz 1963). However, in Stolper's view, economists are not anthropologists or sociologists; moreover, economic planning and economic development are economic problems. While these statements may not in themselves be controversial, what becomes problematic is reducing "national development" to "economic development." The name of the plan, then, is deceptive. It should have been called the "Economic Development Plan" rather than the "National Development Plan." "National development" involves not only economic development but also social and political development. The history of Nigeria since the First Plan (as discussed in Chapter 7) bears testimony to this: Reliance on economic development policies while ignoring sociopolitical development tends eventually to produce the opposite of the intended results.

Despite the above criticism, it should be remembered that at independence in 1960, Nigeria inherited a fiscal situation that posed problems for development. Its unbalanced tax structure depended mainly on foreign trade taxes. In 1960, indirect taxes accounted for 64.2 percent of all revenues, while direct taxes accounted for only 16.5 percent and nontax revenues provided 19.3 percent (World Bank 1976:434). The First Plan, however, failed to provide adequate proposals for improving the tax structure in regard to direct and indirect taxes.

At the end of the 1960s, the national development planning did not seem to have produced the type of political and economic autonomy envisioned by Nigerian leaders and planners. The fact that had all along been ignored is that a politically independent Nigeria was operating within a dependent, colonial economy. The situation resulted in the rise of economic nationalism. As stated earlier, political nationalism led Nigeria to independence in 1960. Economic nationalism, however, resulted from an awareness after political independence that most of the country's economic resources, trade, and commerce were still controlled by foreigners. Nigerian leaders discovered that without economic independence, their "newly won political independence," as Krause (1965:309) puts it, "has little meaning." In this respect, economic nationalism in Nigeria grew out of what Akeredolu-Ale (1975:43) called the "independence effect." Sir Abubakar Tafawa Balewa, the first prime minister of Nigeria, expressed the same view:

> We are anxious to see a large and vigorous private sector developed, but we are also anxious that our own people should take an increasing part in the development of that sector. (Balewa 1964:33)

The federal minister of economic development, Waziri Ibrahim, in his policy statement before the Parliament in 1961, recommended that all foreigners should hand over to Nigerians all distributive trades except large department stores and technical companies (Ogunsheye 1972). As a result of mounting pressure, Kingsley Mbadiwe, federal minister of trade, in 1965 appointed a National Committee on the Nigerianisation of Business Enterprise. Just like a similar committee appointed when he was federal minister of commerce in April 1959 (Advisory Committee on Aids to African Businessmen), its findings were neither implemented nor made public (Balabkins 1982). Other factors that contributed to the delay in the implementation of an indigenization policy included the 1966 military coups and the civil war. The formulation and implementation of an indigenization policy in Nigeria is examined fully in Chapter 7.

SUMMARY: TEN YEARS AFTER INDEPENDENCE

The first decade after independence showed Nigeria's frustration in its efforts to create political unity and autonomous economic development. The period demonstrates why economic development in Nigeria cannot be analyzed in isolation from sociopolitical development. During the 1960s, the chief characteristics of the Nigerian economy were a relatively low revenue effort, heavy dependence on foreign trade taxes, rapidly growing administrative and defense expenditures, low levels of capital expenditures, and relatively slow growth in economic and social services. Also, the regional governments became more dependent upon federal transfers, loans, and withdrawal of funds from marketing board surpluses. Furthermore, increasing current budgetary deficits led the federal government to rely increasingly on internal borrowing to finance its capital expenditure.

Although, from a development perspective, the fiscal performance of the period was not very impressive, it is nevertheless remarkable that a country which inherited a difficult fiscal situation in 1960 and fought a costly civil war from 1967 to 1970 was able to emerge from the decade in decent financial condition. Oil revenues had, however, masked the true effect of war financing. At the end of the 1960s, even though Nigeria was politically independent, its economy remained tied to world market conditions, and in this sense its development continued to be dependent and colonial in nature. However, the oil boom was getting closer. Hope for real development was building up from the increased oil revenues. The political and economic changes that occurred during the oil boom are the subject of Chapter 7.

NOTES

1. Political parties had existed in Nigeria as far back as 1922, when constitutional provisions allowed four electoral seats in the Legislative Council.

2. The role of the military in Nigeria is well covered in Olorunsola (1977) and Odetola (1978).

3. Nyerere's soldiers had mutinied in order to obtain better wages and living conditions, and he was forced to ask for British military intervention. January 24, 1964, was therefore regarded as "the day of national shame" because the independence of Tanganyika was compromised by that request.

4. Others killed during the January 1966 coup included Chief Festus Okotie-Eboh, federal minister of finance; Col. K. Mohammed, chief of staff, Nigerian Army; Lt. Col. A. C. Unegbe, quartermaster general; Lt. Col. A. Largema, commanding officer, Fourth Battalion, Ibadan; Lt. Col. J. T. Pam, adjutant general, Nigerian Army; Col. R. A. Shodeinde, deputy commandant, Nigerian Defence Academy; Ahmed D. Musa, senior assistant secretary (security), Northern regional government; the senior wife of Sir Ahmadu Bello; the wife of Brigadier Ademulegun (Cervenka 1971).

5. According to the Verbatim Report of the Proceeding of Supreme Military Council at Aburi, Ghana, January 4-5, 1967, p. 27, Lt.-Col. Yakubu Gowon was aware of this problem when he stated: "We should keep up the army's oneness as much as possible. Otherwise I think what will happen is that we will start having

private armies; and then you have private police, prisons, and we start dividing the country gradually that way" (quoted in Schwarz 1968:191).

6. Lt.-Col. Odumegwu Ojukwu was quoted in the Verbatim Report of the Proceeding of Supreme Military Council at Aburi, Ghana, January 4–5, 1967, p. 43, as saying: "I do submit that the only realistic form of government today—until tempers have cooled—is such that will move people slightly apart. . . . It is better that we move slightly apart and survive. It is much worse that we move close and perish in the collision" (quoted in Schwarz 1968:191).

7. Politics, no doubt, affected the choices of location of projects, and the selection of contractors, suppliers, and employees (Dean 1972).

8. The determinants of the quality of plan implementation in Nigeria can be grouped under five general headings: (1) the character of the plan and its demands on executive capacity; (2) the adequacy of executive capacity in relation to these demands; (3) the behavior of the donors of foreign aid; (4) the performance of the economy: whether the economy is capable of providing the required resources for implementation; (5) the political system and the interests of politicians (Dean 1972:31–32).

7 Period of Oil Boom and Bust, 1971–85

INTRODUCTION

Chapter 6 established that up to 1970, despite political independence, the Nigerian economy was colonial in nature, dependent on foreign trade and capital attraction for its growth. Up to 1970, there was no clear way to change that condition, and the First National Development Plan certainly did not. Instead, most of the public capital expenditure was invested in infrastructure that supported trade and subsidized private foreign investment. Most private foreign investment in the 1960s went into the petroleum industry, preparing the way for the boom of the 1970s. Therefore, there was little progress economically until the oil boom.

During the 1970s, because of the pool of wealth from petroleum, which was built on trade, leaders and planners thought that development could be transplanted throughout the Nigerian social and economic structures. The country discovered that instead of growth with balanced development, its growth was spectacular, uneven, and, finally, ephemeral.

Between 1970 and 1985, Nigeria was ruled by three military regimes, followed by a democratically elected civilian regime from 1979 to 1983, and then by two military regimes. After a discussion of the formation of OPEC and its impacts on petroleum-producing developing countries (PPDCs), this chapter examines the activities and development strategies of the various Nigerian regimes during the period of oil boom and bust.

OPEC AND ECONOMIC NATIONALISM

In order to understand the role of petroleum in post-independence Nigeria, it is necessary to understand the formation of OPEC and its effects on PPDCs generally.

Juan Pablo Perez Alfonzo of Venezuela, who was instrumental in establishing OPEC, severely criticized the oil companies:

> If there is something absolutely undeniable, as we can plainly see, it is the situation created and maintained in Venezuela by the oil companies. The manner in which they have exploited that wealth belonging to the Venezuelan people, even though aware of the people's needs, is a public and notorious fact. Exploiting the weaknesses of those who, with and without right, have represented the nation, acting beyond the margins of rights and justice, these companies obtained illegitimate profits and caused tremendous ills that cannot be erased by merely introducing a simple legal clause in any law; there does not, and cannot, exist any legal design that can right a wrong. (As quoted in Vallenilla 1975:112)

His thoughts on oil had great impact not only on the petroleum industry in Venezuela, but also in PPDCs. In order for each PPDC to obtain maximum benefits from its oil, Perez Alfonzo recommended nationalization or direct exploitation of hydrocarbons by PPDCs, and the establishment of a common organization for the purpose of raising oil prices and regulating production (Vallenilla 1975).

Venezuela was the first oil producer to start the sharing of profits on a 50–50 basis with the companies. The action resulted in a significant increase in its oil revenues. Venezuela, therefore, would prove to be an important asset to any new petroleum producers' association because of its experience in dealing with the oil companies. Venezuelan interest in an oil-producers' organization began in 1949, when Middle East crudes were showing an economic advantage over those from Venezuela. To rectify the situation, Venezuela would have had to eliminate the profit-sharing system. Instead, Venezuela sent a delegation to Iran, Iraq, Kuwait, and Saudi Arabia to discuss the benefits of obtaining better terms from the oil companies. Persian Gulf state representatives were later invited to visit Venezuela. Thus, the Venezuelan leaders were able to show them the terms they had won from the companies, some of which were also operating in the Middle East. The feeling was that the existing contracts between the companies and the oil states were signed in a colonial atmosphere. Within the next two years, there were significant changes in the terms offered by the companies to the Persian Gulf states. In 1950–51, a system whereby 50 percent of net company profits (based on posted prices) accrued to the government was introduced in Iraq, Kuwait, and Saudi Arabia. In 1953, Saudi Arabia and Iraq entered into the first intergovernment agreement to cooperate in the petroleum market, and in 1959, the first Arab Petroleum Congress, sponsored by the Arab League, was held in Cairo.

The inaugural meeting of OPEC took place in Baghdad, Iraq, September 10–14, 1960. The five original member states at the meeting were Iran, Iraq, Kuwait, Saudi Arabia, and Venezuela. These states came together in Baghdad because they realized that they had to rely to a large extent on petroleum income to balance their annual budgets and to implement national development plans. They felt that petroleum was a "wasting asset and to the extent that it is depleted it must be replaced by other assets" (*OPEC Oil Report* 1977:4).

The principal aims of the participants were to establish consistent petroleum policies to safeguard their interests; to ensure that if any sanction was employed directly or indirectly by any petroleum company against any of the OPEC members, no other member would deal with that company; and to stabilize oil prices. The members felt that they could no longer allow the oil companies to modify prices at will in a fluctuating manner. The repeated price reductions of the late 1950s, implemented by the oil companies without consulting with the oil states, contributed to the formation of OPEC.

OPEC gained strength by admitting new members. Qatar joined in 1961, Indonesia and Libya in 1962, Abu Dhabi (its membership was subsequently transferred to the United Arab Emirates) in 1967, Algeria in 1969, Nigeria in 1971, Ecuador in 1973, and Gabon in 1975. Trinidad and Tobago, Syria, and Congo have applied for membership but have not been accepted. Any oil-exporting country whose petroleum interests are fundamentally similar to those of OPEC founders can become a full member, if accepted by three-fourths of the full members (including the five founding members).

Other events during the latter part of the 1960s aided OPEC's rise to a position of economic power and paved the way for the emergence of a sellers' market in oil. Following the June 1967 Arab-Israeli war, closure of the Suez Canal and Aramco's Tapline to the Mediterranean, and the partial Arab embargo on oil exports, a situation was developing that strengthened OPEC's position in the world energy market. By the early 1970s, the stage was set for rapidly increasing petroleum prices. Saudi Arabian light oil (the "marker" crude) sold at $1.80 per barrel (p/b) in January 1971; by January 11, 1973, it had jumped to $5.12 p/b as a result of an Arab oil embargo set in motion to retaliate for U.S. and European support of Israel. After negotiations failed, the Ministerial Committee of the Gulf Member Countries met in Tehran and set a new posting at $11.651 p/b effective January 1, 1974. Beginning with the winter of 1973–74 "revolution," OPEC determined the price of its petroleum.

Table 7.1 shows the petroleum price revolution of the early 1970s. Oil prices, varying from $1.68 p/b for Kuwait to $2.55 p/b for Libya in January 1971, increased more than 600 percent, reaching $11.55 for Kuwait and $15.77 for Libya in 1974. Hence, the pronounced price increase on January

TABLE 7.1. **Petroleum Price Revolution, 1971–74**
(U.S.$ per barrel)

	Jan. 1, 1971	Jan. 2, 1972	Jan. 1, 1973	Jan. 11, 1973	Jan. 1, 1974
Abu Dhabi	1.88	2.540	2.654	6.045	11.636
Ecuador	n.a.	n.a.	n.a.	10.000	13.700
Indonesia	1.70	2.210	2.260	6.000	10.800
Iran	1.79	2.467	2.579	5.341	11.875
Iraq	1.72	2.451	2.562	5.061	11.672
Kuwait	1.68	2.373	2.482	4.903	11.545
Libya	2.55	3.386	3.777	9.061	15.768
Nigeria	2.42	3.176	3.561	8.171	14.690
Qatar	1.93	2.590	2.705	5.834	12.414
Saudi Arabia	1.80	2.479	2.591	5.119	11.650
Venezuela	2.12	2.990	3.160	7.240	14.080

Source: Vallenilla 1975:172.

1, 1974, was viewed as an oil crisis by the industrial nations. However, these increases were seen by OPEC members as "guided" market adjustments necessary to compensate them for the many years in which their energy resource had been sold at token prices. The high profit figures reported by oil companies after the crude oil price increases contributed to the $0.38 p/b increase established by OPEC at the December 13, 1974, conference in Vienna, which price was to take effect January 1, 1975. These price increases provided the PPDCs with undreamed-of income, so vast as to cause enormous changes in the role of oil in their economies.

The rest of this chapter is devoted to the discussion of Nigeria's economic and sociopolitical changes during the 1970s. First, the role of oil revenues in the economy during the 1970s is summarized, and the various Nigerian regimes and their strategies for development during the oil boom are examined.

PETROLEUM AND THE GENERAL ECONOMY

Before embarking on a discussion of the various regimes that governed Nigeria during the period of oil boom, it is advantageous to become familiar with the economic conditions under which those regimes operated. The aim is to show how the oil revenues provided an enormous amount of capital and foreign reserves, so that what is usually cited as the main constraint on development in developing countries—lack of capital and foreign exchange—ceased to be a constraint in Nigeria during the 1970s. Then an analysis of the economic changes that occurred during the period of growing petroleum revenues is provided.

Table 7.2 presents Nigeria's GDP between 1970 and 1982. In 1958, the first year of petroleum shipment from Nigeria, the GDP was N-1.88 billion. By 1960, it grew to N-2.4 billion, and to N-3.36 by 1965. By 1970, despite the damages from the civil war, the GDP had increased to N-5.62 billion, and within four years it had skyrocketed to N-16.96 billion. By 1980, GDP had grown by over 255 percent, to N-43.28 billion. It is important to note that the phenomenal growth recorded during the 1970s stopped in 1981 with the decline in petroleum demand.

TABLE 7.2.   Gross Domestic Product, 1950–82

Year	Million naira
1950	524
1958	1,880
1960	2,400
1965	3,361
1970	5,621
1971	7,098
1972	7,703
1973	9,001
1974	16,962
1975	20,405
1976	25,449
1977	28,015
1978	33,738
1979	39,939
1980	43,280
1981	43,450
1982	44,884

Source: IMF 1984: 454–55.

Table 7.3 contains crude oil production figures for Nigeria between 1970 and 1984. The rapid increase in GDP during the 1970s was attributed to the phenomenal increase in oil production. Average liftings increased from 0.174 million b/d in 1960 and 1.083 million b/d in 1970 to 2.25 million b/d in 1974. Then world recession in the first half of 1975 pushed production back to 1.79 million b/d. Production rose again in 1976 and 1977, exceeding the 2 million b/d mark in both years before declining in 1978 to 1.9 million b/d. In 1979, production increased to 2.3 million b/d, and since 1980 has declined continuously. In 1981, for example, production fell drastically—as low as 0.6 million b/d, even though the year ended with an average of 1.43 million b/d. As part of OPEC's strategy to control supply, Nigeria in 1984 maintained an OPEC-imposed quota of 1.3 million b/d.

TABLE 7.3.   Petroleum Production, 1970–84
(million b/d)

Year	Production
1970	1.083
1971	1.531
1972	1.816
1973	2.057
1974	2.254
1975	1.786
1976	2.077
1977	2.097
1978	1.908
1979	2.301
1980	2.065
1981	1.430
1982	1.289
1983	1.235
1984	1.300*

*OPEC-imposed quota.
Sources: OPEC Oil Report 1979:258; FRN 1984b; Africa Guide 1983:276, 278.

No doubt the country benefited greatly from the oil price revolution of the 1970s and the early 1980s. Prices rose from U.S. $2.25 p/b (per barrel) in 1970 to U.S. $14.56 in 1977. In line with OPEC pricing policy, Nigeria increased its oil prices several times during 1979 and 1980. By March 1982, it had reached U.S. $36.50, then was reduced to $30 p/b in February 1983, and to U.S. $28 p/b on October 18, 1984, in response to similar cuts by two non-OPEC competitors, Britain and Norway. The higher prices of the late 1970s and the early 1980s had masked the true position of Nigeria's single-product, export-based economy: that demand was declining even though the value of exports skyrocketed from ₦6.1 billion in 1978 to ₦10.68 billion in 1979 and ₦14.6 billion in 1980. In line with declining export of petroleum since 1981, export earnings have continued to fall. Table 7.4 provides the figures for oil as a percent of total export earnings.

Petroleum is Nigeria's main source of foreign exchange, contributing over 90 percent of the total. In 1980, petroleum exports reached ₦14 billion. Correspondingly, Nigeria's foreign reserves grew and declined with petroleum exports (see Table 7.5).

TABLE 7.4.   Oil Exports as Percent of Total Exports, 1950–83
(million naira)

Year	Total Exports	Petroleum	Percent of Total
1950	180	0	0.00
1958	271	2	0.74
1960	339	9	2.65
1965	529	131	24.76
1970	886	510	57.56
1971	1,304	964	73.93
1972	1,433	1,175	82.00
1973	2,319	1,935	83.44
1974	6,104	5,675	92.97
1975	4,791	4,592	95.85
1976	6,322	5,895	93.25
1977	7,594	7,046	92.78
1978	6,707	6,033	90.00
1979	10,676	10,035	94.00
1980	14,640	13,999	95.62
1981	11,892	11,250	94.60
1982	11,145	10,503	94.24
1983	8,427	7,786	92.39

Source: IMF 1984:454–55.

TABLE 7.5. Foreign Reserves, 1950–83

Year	U.S.$ Million
1950	110
1958	505
1960	343
1966	195
1970	202
1971	408
1972	355
1973	559
1974	5,602
1975	5,586
1976	5,180
1977	4,232
1978	1,887
1979	5,548
1980	10,235
1981	3,895
1982	1,613
1983	990

Source: IMF 1984:452–53.

In 1958, the first year of oil shipment, Nigeria's foreign reserves were ₦ 505 million, mainly from agricultural exports. By 1970, at the end of the civil war, reserves stood at ₦ 202 million. Three years later, reserves had more than doubled to ₦ 559 million. In 1974, reserves grew to ₦ 5,602 million. Since then it has fluctuated as low as ₦ 1,887 million in 1978 and as high as ₦ 10,235 million in 1980.

Considerable changes, resulting from the oil boom of the 1970s, can be seen in the various sectors of the Nigerian economy (see Table 7.6). During 1971–77, the construction sector enjoyed an average annual growth rate of 24.7 percent, up from 9.0 percent during 1960–70. Public administration and defense grew at an average annual rate of 24.6 percent—up 4.8 percent from 1960–70. Transport and communication grew from an average annual rate of 4.0 percent during 1960–70 to 16.5 percent during 1970–77. Electricity, gas, and water rose from 16.7 percent to 18.2 percent during the same periods. Manufacturing grew from 12.8 percent to 13.4 percent. The mining sector decreased from an average rate of 31.7 percent during the 1960s to 8.1 percent in 1970–77.

Agriculture, which grew at an average annual rate of only 1.3 percent during the 1960s, did not grow during the 1970s and, in fact, recorded a

TABLE 7.6. Sectoral Average Annual Growth Rate, 1960–77

Sector	1960–70	1971–77
Agriculture	1.3	-1.5
Mining	31.7	8.1
Manufacturing	12.8	13.4
Construction	9.0	24.7
Electricity, gas, water	16.7	18.2
Transport, communication	4.0	16.5
Public administration, defense	4.8	24.6
Imports of goods and services	4.0	24.7
Exports of goods and services	12.9	5.6
GDP at factor cost	4.3	6.2
GDP at market prices	4.4	6.0

Source: World Bank, 1980:150–51.

negative annual growth rate of 1.5 percent. Agricultural products constituted over 80 percent of total exports in 1960; by 1970, 44 percent; and by
1978–79, only 6 percent of the total (*Africa South of the Sahara* 1983:639).
No doubt the phenomenal growth of the petroleum sector contributed to
this decline. There were other reasons for the decline, however: drought,
low producer prices, and competition for labor from the other economic
sectors. Before the oil boom, the main agricultural export products included cocoa, peanuts and peanut oil, cotton, palm kernels, rubber, and
timber—all of which had been developed in response to colonial needs.
At present, cocoa is the only significant agricultural export.

Exports of goods and services, which had increased by 12.9 percent
during 1960–70, when agriculture was still growing and the mining sector
was increasing at an average annual rate of 31.7 percent, declined to 5.6
percent during 1970–77, as mining and agriculture declined. Imports experienced a dramatic average annual growth rate of 24.7 percent during
1970–77, up from only 4.0 percent during 1960–70. It should be noted,
however, that imports increased correspondingly with construction,
public administration and defense, and transport/communication, but
not with manufacturing. Table 7.7 provides the exports/imports ratio for
Nigeria during the 1970s and the early 1980s. What had happened was

that cash cropping and petroleum production for export had restricted Nigeria to the role of a primary producer while foreign manufacturers rushed mass-produced manufactures at exploitative prices to the country.

TABLE 7.7. Exports and Imports, 1970–82

Year	Exports (E)	Imports (I)	Ratio E/I*
1970	954	937	1.018
1971	1,422	1,328	1.070
1972	1,522	1,286	1.184
1973	2,467	1,808	1.364
1974	6,244	2,743	2.276
1975	5,453	4,988	1.093
1976	7,840	7,074	1.108
1977	8,481	8,787	0.965
1978	7,182	9,083	0.791
1979	10,712	8,903	1.203
1980	13,494	11,004	1.226
1981	11,649	14,969	0.778
1982	9,395	10,817	0.869

*Author's calculation

Source: IMF 1984: 454–55.

With these conditions in the 1970s, Nigeria entered the euphoric boom years and what could be called the years of planning for economic growth with surplus capital. Also, from a conventional economic point of view, the shifting, during the 1970s, from agriculture to mining, manufacturing, construction, and other sectors could be read as Nigeria's entry into a modern industrial age.

With the above brief analysis of the Nigerian economy during the oil boom, we can now proceed to examine the Nigerian regimes of the 1970s and the early 1980s.

THE GOWON REGIME: 1970–75

The 1970s was a period of change both in the economy and in the sociopolitical structure of Nigeria. It was the period of oil boom and bust. It was also during that decade that democracy was restored in Nigeria after 14 years of military rule.

The civil war in Nigeria, which lasted for 30 months, ended on January 12, 1970. Called the War of National Unity, it was believed to have settled the question of "One Nigeria," and that secession was therefore out of the question. Following the war, however, the Nigerian government found itself faced with two problems during the 1970s: the task of postwar reconstruction (physical, social, economic, and moral) and a program for the return to civilian rule.

On October 1, 1970, Gen. Yakubu Gowon, in a State of the Nation policy broadcast, set out a nine-point program on which, according to him, lasting peace and political stability in Nigeria could be built (Kirk-Greene and Rimmer 1981:4). These included the reorganization of the armed forces; the implementation of the National Development Plan, and the repair of the damage and neglect of war; the eradication of corruption from Nigeria's national life; the settlement of the question of the creation of more states; the preparation and adoption of a new constitution; the introduction of a new formula for revenue allocation; the conducting of a national population census; the organization of genuinely national political parties; and the organization of elections in the states and the nation.

The target date for completing these tasks was set for 1976—ten years from the beginning of military rule in Nigeria. The announcement came as a shock to most Nigerians. In April 1967, Gowon had envisaged a two-year period before return to civilian rule, but with the secession of the former Eastern Region on May 30, 1967, and the subsequent civil war, return to civilian rule in 1969 became unrealistic. Still, in 1970 most Nigerians could not understand why it would take six more years to return to a democratically elected government.

In executing its massive program of postwar physical reconstruction and the ambitious 1970–74 National Development Plan, Nigeria benefited greatly from its new oil wealth. In 1960, oil export represented only 2.65 percent of all exports; by 1970, it represented 57.56 percent. A year later, it represented 73.93 percent, and by 1974, its share of all exports had jumped to 92.97 percent (see Table 7.4). Also, in 1970, petroleum production was 1.1 million b/d; by 1974, it had grown to 2.6 million b/d (see Table 7.3). In 1970, Nigeria's foreign reserves stood at U.S.$0.2 billion; by 1974, they had increased to U.S.$5.6 billion (see Table 7.5). For this reason, the Gowon regime could be said to be at the right place at the right time. The economy was booming from the oil revenues to an extent undreamed of during the 1960s. Nigeria's GDP, which stood at N 3,361 million in 1965, increased to N 5,621 million in 1970. By 1974, it had increased to N 16,962 million and to N 20,405 million by 1975 (see Table 7.2).

According to the Guideposts for Second National Development Plan, the most important aim of the plan was

> a high overall rate of economic growth with a view to achieving self-sustained growth . . . rapid industrialisation of the economy; increased production of food for domestic consumption without relaxing efforts in the export sector; and a drastic reduction in the magnitude of the present unemployment problem. (FRN 1966:2–3)

The emphasis of the Second Plan was on the need to achieve the highest possible growth rate of per capita income (FRN 1970:34). It also mentioned other objectives that may be classified as national autonomy and social justice. The need to achieve national autonomy, according to the Second Plan, would be reflected in Nigeria's foreign policy. This autonomy was for Africa and for Nigerian ownership of enterprises. As stated in the plan, "The uncompromising objective of a rising economic prosperity in Nigeria is the economic independence of the nation and the defeat of neo-colonialist forces in Africa." Also, in order to foil the "global strategy of modern international combines," the Nigerian government would progressively sustitute Nigerian for foreign ownership and management of economic enterprises (FRN 1970:31–34). This would take the form of indigenization and Nigerianization (FRN 1970:144, 289).

The Indigenisation Decree of 1972 was a direct result of this aim. According to the plan, "a Government cannot Plan effectively what it does not control"; Nigerian development planners concluded that the federal government must "acquire and control on behalf of the Nigerian society, the greater proportion of the productive assets of the country" by participating in the equity ownership of private enterprises or by exclusive public ownership. The exploitation of "strategic national resources" was to be reserved for the federal govrenment. The federal government, according to the plan, must play a leading role in mining and manufacturing, which were considered as important revenue sources for future development (FRN 1970:33, 289). We shall return to the question of economic nationalism later in this section.

The plan also was intended to create a just and egalitarian society, providing a more equitable distribution of income among persons and promoting balanced development among the various communities (FRN 1966:3; FRN 1970:33–34, 71–72). As far as the equity question was concerned, the progress reports on the implementation later admitted that lack of adequate information on income distribution prevented attempts to devise an effective policy on incomes (FRN n.d.:35; FRN 1974:32).

An examination of the percentage distribution of public sector capital investment by sectors during the Second Plan period (1970–74) shows that general administration, defense, and security topped the list with 22.7 percent. Second in order of priority was transport with 21.3 percent. Education received 11 percent, followed by agriculture (including livestock, forestry, and fishery) with 9.7 percent; and trade and industry, 7.3 percent. The other sectors' shares were as follows: water (other than irrigation), 5.8 percent; electricity, 5 percent; health, 5 percent; communication, 2.4 percent; town and country planning (including housing), 2.2 percent; financial obligations, 2.1 percent; labor, social welfare, and sports, 1.3 percent; and information, 0.9 percent (FRN 1970).

The Second Plan was to serve as a means of reconstructing the damage caused by the civil war and, at the same time, to promote economic and social development throughout the country. The sectoral distribution of capital expenditure, as shown above, reflects this consideration. The phenomenal increase in the size of the armed forces during the civil war resulted in a sharp increase in the proportion of capital expenditure allocated to general administration, defense, and security. It rose from 7.1 percent in the First Plan (1962–68) to 22.7 percent in the Second Plan. Transport, which was number one in the First Plan, dropped to number two but maintained the same proportion of capital expenditure—21.3 percent during both plans. A distant third was education with 11 percent, an increase of 0.7 percent over the First Plan. Agriculture not only dropped from third place in the First Plan to fourth place in the Second Plan, but its proportion declined from 13.6 percent to 9.7 percent. Trade and industry (including mining) suffered a similar decline: from 13.4 percent during the First Plan to 7.3 percent in the Second Plan. The urban bias of the plan becomes clear when one considers that only 18.2 percent of all planned investments went to the rural areas while urban investments accounted for 81.8 percent (Sada 1981:276–77). Other biases can be found, such as the high proportion of administrative projects in relation to industrial and agricultural ones.[1]

On the other objective, which involved national autonomy, the 1970–74 plan provided steps for restricting the activities of foreigners in certain sectors of industry and commerce, and emphasized the reduction in the share of capital owned by foreigners. Its goal was to increase indigenous participation in the ownership of businesses and to increase the placement of Nigerians in management and professional jobs.

Before embarking on a discussion of the indigenization process in Nigeria, it may be appropriate to define indigenization, Nigerianization, Africanization, and expropriation. Indigenization refers to the substitution of private indigenous Nigerian ownership for foreign ownership and control of the modern economic sector of the country. Nigerianization, on the other hand, means replacing foreign senior personnel with Nigerians, both in the public and in the private sectors. Africanization means the substitution of African for non-African ownership and control. Expropriation refers to the right of a government to take over private property, usually (but not always) with compensation to the owners, who usually are foreigners. The economic transformation that took place in Nigeria did not include expropriation such as occurred in Uganda or Ghana; it took the form of buying out foreign businesses and forming indigenous capital in Nigeria.[2]

Like all the other developing nations, Nigeria is aware of the inequality of economic and political strength among the nations, and that the few rich and dominant nations determine the pattern of international trade, terms of trade, technology transfer, direct investment, and foreign aid. Nigeria has experienced the transfer of inappropriate technology and skills, and the dumping of exports on its markets. These conditions create a feeling of economic dependence and economic vulnerability among Nigerians. Therefore, after the civil war, it was time to turn the country's attention to the task of breaking the shackle of colonial dependence. There was a need for a decree that would transfer the control and management of foreign-owned and foreign-managed enterprises to private sector Nigerians. The entire country, including the press, was united on the subject. Even the *Nigerian Observer* (of February 27, 1974) compared foreign business ownership in Nigeria to "a rotten tooth resting comfortably in the nation's mouth" that must be extracted. The feeling was that it was better to mismanage one's economy than for foreigners to retain a lien on it (Uzamere 1974). Also, during the civil war, foreigners controlled 95 percent of Nigeria's large-scale and 82 percent of the medium-scale industries; these facts led to the promulgation of the Indigenisation Decree of 1972 (Aluko 1972).

The Indigenisation Decree of 1972, which actually was the "Nigerian Enterprises Promotion Decree of 1972 (Decree No. 4)," was signed by General Gowon on February 23, 1972. It was to become effective on March 31, 1974—"The Appointed Day" (FRN 1972). Its purpose was to remove foreign domination of commerce and industry in Nigeria, and to give the people "the right and power to opt out of the demoralising dependence syndrome and take the destiny of their nation into their own hands" (Tukur and Olagunju 1973:114). Indigenization, it was believed, would result in "maximum retention of profits at home" (Bello 1975:8). It was believed that the less profits were repatriated abroad, the better the country's balance of payments was likely to be. To achieve these goals, in the trade, industrial, and service sectors a compulsory transfer of foreign ownership to Nigerians was necessary. It was not the intention of the decree to nationalize the economy or to expropriate the economy.

The decree contained two parts referred to as Schedule I and Schedule II. Schedule I declared 22 industrial and service activities to be exclusively for Nigerian citizens or associations. These included advertising and public relations agencies, all aspects of pool betting and lotteries, casinos, blending and bottling of alcoholic drinks, tire retreading, haulage of goods by road, bus service and taxis, and assembly of radios, television sets, and other domestic products.

Schedule II listed 33 commercial and industrial activities that also were reserved for Nigerians but that, under certain conditions, could be owned and managed by foreigners, subject to a minimum Nigerian equity participation of 40 percent. Some of the activities were coastal and inland waterways shipping, construction, department stores and supermarkets, distribution agencies for machines and technical equipment, distribution and servicing of motor vehicles, tractors and spare parts, domestic air transport (scheduled and chartered), meat processing, beer brewing, manufacture of cement, manufacture of metal containers, and boatbuilding. Foreign firms were classified into either schedule on the basis of tax accounting information submitted to the Federal Board of Inland Revenue for the fiscal year 1968–69, 1969–70, and 1970–71.

The principal organ for implementing the Indigenisation Decree of 1972 was the Nigerian Enterprises Promotion Board (NEPB), an arm of the Federal Ministry of Industries. To deter nonconformists after March 31, 1974, the decree empowered the NEPB to impose fines and dispose of the defaulter's property in addition to applicable prison terms. The 1972 Indigenisation Decree also set up 12 advisory Nigerian Enterprises Promotion Committees, one in each state. By 1975, there were optimistic reports showing a 77.5 percent compliance rate among the businesses targeted for indigenization (Ekukinam 1975:2).

In 1973, while the hope for economic autonomy was still running high, Gowon decided to implement the seventh point of his nine-point program: conducting a national population census. A population census had become a very sensitive issue in Nigeria for several reasons. First, political power hinged upon it; if and when a civilian government was created, the number of seats allocated to each region in the federal Parliament would be based on population. Second, population was one of the factors used in distributing the national revenues to the regions (and later, the states). Third, the 1963 census had left an unpleasant memory in the minds of most Nigerians. It reminded them of the saying that the First Republic of Nigeria was a country that "had censuses that were not censuses, elections that were not elections, and finally governments that were not governments (*Nigerian Opinion* 1966).

Gowon was aware of the 1963 episode and was determined to avoid a repeat. But, contrary to his expectations, the 1973 census results and the reactions to them were worse than those of 1963. According to the results, the population of Nigeria had grown from 55 million to 79 million within a decade. Second, the population figures of all northern states showed growths ranging from 29 percent to 97 percent. For the southern states, with the exception of Lagos, which gained 72 percent in population, population growths ranged from 11 percent to 45 percent. The Western State, which was not a battlefield of the civil war, had lost 500,000 people between 1963 and 1973 (6 percent of its 1963 population), while the South-Eastern State lost 4 percent. Hence, the census rekindled Nigeria's ethnic tensions and fears—some groups again began to see traces of perpetual ethnic domination in the structures of the Nigerian economic and political systems.

In responding to the reactions to the 1973 census figures, General Gowon, in a national broadcast on October 1, 1974, announced his decision to postpone indefinitely the return to a democratically elected civilian rule. He felt that moderation and self-control were still lacking among Nigerians. While it is possible to defend Gowon's decision as being in the best interest of the country at that time, he had been criticized for his failure to implement any of the other measures he deemed necessary before a return to civil rule. These measures include the redeployment of the 12 state military governors—most of whom had been in office for seven years; the appointment of a new set of federal commissioners; the establishment of a panel to draw up a new constitution; and the setting up of another panel to settle the question of creation of more states. Added to all these was his ineptness in handling the Udoji Report on the civil service, which was presented to the government in September 1974. That report awarded a huge payment of salary arrears to public employees while excluding private sector employees. The inflationary effect of such salary increases, and the impacts on the cost of living of those not eligible, led to nationwide industrial unrest. In addition, there was a mysterious shortage of gasoline throughout the country. Most Nigerians began to believe that the national reconstruction program had gone off track.

On July 27, 1975, Gowon flew to Kampala, Uganda, to attend the Organization of African Unity summit conference. Two days later, he was overthrown in a bloodless coup. It was exactly nine years after Gowon took power in Nigeria's second military coup.

In assessing the success of Gowon's nine-point plan, Arnold (1977:23-24) had this to say:

> Of these nine points five got nowhere. Corruption surely got worse; the question of more states was ignored, as was the preparation of a new constitution; while the organization of national political parties and elections for popular governments were set aside by Gowon in his 1 October 1974 speech. Of the rest: the census, perhaps not surprisingly in terms of

Nigeria's history, was so fraught with explosive political possibilities that it was the first thing the Mohammed regime cancelled on coming to power. The Second Plan—one way and another—was partially fulfilled, the failure here being due mainly to lack of sufficient executive capacity. The army was supposedly reorganized, though not according to Gowon's successors. A new and on the whole sound system of revenue reallocation was introduced. But in terms of putting into operation a political programme it was hardly an exciting or successful record.

Despite the above assessments, a more precise evaluation of Gowon's regime would show that up to July 1974, he was, in the opinion of most Nigerians, "the good young Jack," the God-fearing leader. His fall stemmed from his taking the goodwill of Nigerians for granted. His broken promises about restoring civilian rule in 1976, the creation of states, and the reassignment of state governors dissipated the support he had among most Nigerians. But he will be remembered for his handling of the civil war. Under his regime, the economy did not achieve the type of autonomy that was envisioned. It remained colonial in nature, became more dependent on oil, and more geared toward international trade and capital accumulation. When Gowon left office, despite reported corruption and waste at the federal and state levels, he managed to leave behind foreign reserves in excess of ₦5 billion—up from ₦0.2 billion in 1970. His nine-point plan, even though not fully executed by Gowon, nonetheless formed a basis for the next administration's strategies, as discussed below.[3]

THE MURTALA MOHAMMED/OLUSEGUN OBASANJO REGIME: JULY 1975–OCTOBER 1979

On Tuesday, 29 July 1975, two days after Gowon left Lagos to attend the OAU summit meeting of heads of states, Nigerians awoke to learn that the government had been overthrown. The new head of state, Brig. Murtala Mohammed, had served under Gowon as federal commissioner for communications. Born in 1938, he received most of his military training in England and was one of the most successful commanders during the civil war.

In a national broadcast the following evening, Mohammed stated the reasons why Gowon was removed from office. These included an inability to meet the aspirations of the people despite the resources available to the country; his inaccessibility to the people; the indecision and indiscipline of his administration; and his neglect of the administration of the armed forces. He announced that the new government would rule through three organs: the Supreme Military Council, the National Council of States, and the Federal Executive Council. Under the new government, state governors were to be responsible, through the chief of staff, Supreme Headquarters (Brig. Olusegun Obasanjo), to the head of state. The controversial 1973 census was canceled, and the 1963 census was to be used for planning purposes. Meanwhile, a commission was to be set up to review the question of new states (in addition to the existing 12 states), and another commission to advise the government on the question of the federal capital.[4]

Throughout the country, there was a general acceptance of the coup; people felt that the Gowon regime had failed the nation and a change was inevitable. The replacement of the 12 state governors and the removal of state governors from the Supreme Military Council was well received by the people. Over the next several weeks, there was a major purge of the bureaucracy. "Operation Deadwoods," as it was called, extended to the entire country. By the time the operation was over at the end of November 1975, over 10,000 public employees had been retired (with or without pension rights) or dismissed (Kirk-Greene and Rimmer 1981).

Despite these changes, the return to democratic rule was of utmost concern to Nigerians. In his broadcast to the nation on National Day, October 1, 1975, Mohammed reiterated the military government's intention to hand power over to a democratically elected government on October 1, 1979. A five-stage program was to be implemented before that target date. First, to resolve the issue of the creation of new states, final decisions were to be made by April 1976. Also, the Constitution Drafting Committee was given 12 months to complete its task. During the second stage, the newly created states would have two years to put their governments in place. The local government system would be reorganized, with elections at the local government level based on individual merit, without political parties. From that level, a Constituent Assembly would be organized, partly elected and partly appointed. That body would consider, amend, and finally approve the draft constitution. The deadline set for completing the second stage was October 1978, so as to facilitate the preparation for the general elections. In stage four, the elections for the state legislatures would take place. Stage five covered the elections at the federal level. These last three stages would extend over one year, so that the military could withdraw by October 1, 1979.

Up to 1976, there were major differences in the system of local government in Nigeria, and even in the names by which local governments were known. There were Administrative Areas, Local Authorities, Development Areas, Divisional Authorities, and Local Administrations, and the Western State Council-Manager system. In Chapter 4, it was brought out that the British colonial government was committed to the principle of indirect rule and that the Native Authority system was implemented more successfully in the Northern Region of Nigeria, where the apparatus was already in place in the emirates. Oyediran and Gboyega (1979:169) described local government systems in Nigeria and expressed the need for uniformity:

> When the military seized political power in 1966. . . . there was no local government worthy of the name in Southern Nigeria. In the Northern Region, on the other hand, local government (the system of Native Authorities) was very strong indeed, and the emirs and traditional rulers who wielded its instruments were in name and deed the "government" in their localities. In practical terms, in Southern Nigeria the need was to revitalise and consolidate the local government system: the local government has become the object of contempt, perhaps pity as well, but never support. In Northern Region the need was to modernise the system by emphasising popular participation and control.

Because of these differences in systems, the federal military government enacted the States Local Government Edict of 1976. The edict requires all councils to have a majority of members elected to formulate policy by a majority vote, and to elect their own chairpersons. The new local governments were a single-tier authority and a third level in the federal system of government.

The panel appointed to look into the issue of more states presented its report at the end of 1975. Shortly thereafter, General Mohammed announced the creation of 7 additional states, bringing the total to 19. He urged the increase to 19 to be accepted as the final disposition of the agitation for new states.

On Friday, February 13, 1976, while General Mohammed was driving to his office in Lagos, he was assassinated in an attempted coup. One military governor, Col. Ibrahim Taiwo of Kwara State, was also killed. The coup was led by Lieutenant Colonel B. S. Dimka. At the time, there were suggestions that foreign powers were involved in the coup (Arnold 1977). On the Sunday after the coup, there was a student demonstration at the American Embassy. The demonstrators carried placards implying a link between Dimka and the CIA. The coup was also linked, by implication, to Britain. On the day of the coup, Dimka and two armed soldiers went to the British High Commission in an attempt to telephone General Gowon, who since his overthrow had been studying in Britain (*Dimka's Confession,* 1976:5, 40). The British High Commission refused to allow this or to forward a message to Gowon. Dimka's visit to the High Commission led to a students' attack on the High Commission the following Tuesday. Shortly thereafter, the British high commissioner, Sir Martin LeQuesne, was withdrawn from Lagos at Nigeria's request. Among 32 Nigerians implicated in the coup and executed by firing squad was the former defense commissioner, Maj. Gen. I.D. Bisalla.

The chief of staff, Supreme Headquarters, Gen. Olusegun Obasanjo, who was the next ranking officer, assumed control of the government. He regarded his regime as a continuation of Mohammed's, and sought to carry out the rest of the five-stage program for handing over power to a democratically elected civilian government.

The new local governments were used as electoral colleges for the selection of Constituent Assembly (CA) members on August 31, 1977. Allocation of membership was based on estimated population. Of the 230 members, 20 were nominated by the Supreme Military Council. During the eight months of constitutional debate (October 6, 1977–June 5, 1978), young radicals had an opportunity advocate formal socialism, while the majority of the CA believed that socialism was unsuitable for Nigeria. Also, the major political parties that took part in the 1979 election were formed among the various informal groups of the CA.

After the revised draft of the new constitution was presented to General Obasanjo on August 29, 1978, his government introduced 17 amendments before its promulgation three weeks later. The constitution was seen as a prerequisite to a democratically elected government in Nigeria. It was an instrument aimed at furthering political development, which was expected to accompany economic development. But the new constitution left unresolved two important questions relating to economic development: the census and the appropriate formula for revenue allocation. They were left, because of their political nature, to the incoming civilian administration.

The ban on political parties was lifted by General Obasanjo on September 21, 1978, so that preparation could start for stages four and five of the Mohammed Plan. In those two stages, elections to the state legislatures and then to the federal National Assembly (the House of Representatives and the Senate) were held. The severe compression of time for registering new political parties led to the formation of many new but hurriedly constituted parties. By the end of November 1978, the number had reached 50. It was left to the Federal Elections Commission (FEDECO) to enforce the provisions of the Electoral Decree of 1977 by scrutinizing all the political associations and to register as political parties only those meeting stipulated conditions—for instance, the name of the association, its aims, its emblem, its motto, its membership, and the location of its headquarters must demonstrate that the association was a national party rather than an ethnic or regional or state party. Also, its executive committee was required to reflect the federal character of Nigeria. Of the 50 parties reportedly formed, 19 filed papers with FEDECO by the mid-December 1978 deadline. Five parties were found by FEDECO to have met the conditions of the Electoral Decree (Diamond 1983). All these parties, in one way or another, had links with the political parties of the First Republic.

The Unity Party of Nigeria (UPN), led by Chief Obafemi Awolowo, leader of the old Action Group (AG), who also served in the Gowon regime as federal commissioner for finance, offered a program of welfare socialism emphasizing free health care and free education at all levels. A second party, the National Party of Nigeria (NPN), was launched under the patronage of Alhaji Aliyu, Makaman Bida, one of the founders of the old Northern People's Congress (NPC), which had controlled the federal government up to 1966, when the army took over. In its election manifesto, the party announced its program, which included the security of the state and of life and property, guaranteed by the provision of adequate means of external defense, and internal law and order; granting full respect and recognition to the traditional rulers; and the encouragement, protection, and inducement of foreign capital and technology (NPN 1979). Alhaji Shehu Shagari was nominated as its presidential candidate.

The People's Redemption Party (PRP) was led by Alhaji Aminu Kano, who in 1951 had broken away from the NPC to form the Northern Elements Progressive Union (NEPU). His party, committed to social revolution, denounced exploitation, oppression, and feudalism. It offered a program of "democratic socialism" that sought self-reliance and the transfer of the economy from the grip of neocolonialists to the hands of the people (PRP 1979). Aminu Kano had held high political offices under the First Republic and under the Gowon regime.

A fourth major party was the Nigerian People's Party (NPP), founded by Alhaji Waziri Ibrahim, a millionaire businessman from Borno who had held ministerial appointments under the First Republic. When his nomination for the presidency was opposed by Nnamdi Azikiwe and Chief Adeniran Ogunsanya, and in essence he lost the leadership of the party, he formed the Great Nigerian People's Party (GNPP), which became the fifth major party. He was its presidential candidate. Azikiwe was nominated as the presidential candidate of NPP. Both the NPP and GNPP espoused social justice and social change, just as the UPN did.

Thus, by early 1979 the stage was set for a return to democratically elected government. The elections for the presidency, the federal Senate, federal House of Representatives, the state senates, and state houses of representatives were scheduled for five Saturdays during July and August 1979. The first four were held one week apart to allow sufficient time for the announcement of one result before the next election took place. The final one, the presidential election, was scheduled two weeks after the election of the state houses.

In contrast with the 1964–65 elections, the 1979 elections (which employed nearly 500,000 electoral staff to administer a voter registration roll of 48 million people in 125,000 polling booths) was conducted without serious incident. In fact, the election was marked by apathy—for the presidential election, only 16.8 million out of 47.7 million registered voters actually cast their ballots (Kirk-Greene & Rimmer 1981). It is possible, however, that the people felt overpolled because the elections were spread over six weeks. All the results pointed to the NPN as the majority party, and Alhaji Shehu Shagari was declared the president-elect of the Second Republic. He was installed as president of Nigeria on October 1, 1979.

During the Mohammed/Obasanjo regime, the Third National Development Plan 1975–80 (hereafter referred to as the Third Plan) was implemented. It was larger and more ambitious than the Second Plan. Its objectives included an increase in income, a more even distribution of income, a reduction in unemployment, balanced (geographically dispersed) development, a diversification and indigenization of economic activities, and an increase in the supply of high-level manpower (FRN 1975a:27). The strategy of the Third Plan was the internalization of the rapid growth of the oil sector. That meant finding the best ways to utilize the increased revenues from oil exports, to create the infrastructure of "self-sustaining growth." In 1975, petroleum production was 1.8 million b/d (a decline from 2.2 million b/d in 1974). By 1979, production had recovered to 2.3 million b/d. In 1975, petroleum exports accounted for over 95 percent of all exports, and in 1979, for 94 percent. Petroleum exports increased from ₦ 4.6 billion in 1975 to ₦ 10.04 billion in 1979.

As a way of altering the distribution of income, the poorer areas of the country would receive subsidized public services such as electricity, water supplies, health services, cooperatives, and community development. Free primary education throughout the country was introduced (FRN 1975a:28–29, 46). Some of the other measures included the reduction in secondary school tuition, board, and lodging fees; free education at all universities, teacher training colleges, colleges of education, and colleges of technology; and, from April 1, 1979, abolition of tuition fees in all secondary schools.

Also, measures were taken to raise agricultural productivity and improve the income of farmers. The state marketing boards had come under severe criticism for their pricing policy, which had a negative effect on agricultural production. When they were created in the late 1940s, their objective was to stabilize local prices over the year. Instead, it became to raise revenues for development by withholding funds from farmers. To correct the situation, in 1973 the federal government took over the power to fix export produce prices from the state marketing boards. The federal government replaced the export duties and states' produce sales taxes with a single federal tax of 10 percent, which was removed in 1974. The goal was to maintain higher producer prices, and thus reduce "the degree of income inequality" (FRN 1975a:14). In 1977, the state marketing boards were replaced with a national board for each commodity.

The 1975–80 Plan originally called for a total investment of ₦ 30 billion. An examination of the Third Plan shows that trade and industry (including mining) received the largest share (26 percent), an increase from 7.3 percent in the Second Plan (1970–74). The transport sector's allocation was 22.2 percent, a slight increase over its 21.3 percent in the Second Plan. General administration including defense received 13.6 percent, down from its 22.7 percent share of the Second Plan; town and country planning, 9.2 percent, increasing by over 400 percent over its allocation in the Second Plan. Education received 7.5 percent, down 32 percent from its 11 percent share in the Second Plan. Communication almost doubled its 1970–74 share—it received 4.1 percent. Allocations to the other sectors, such as electricity, water, and health, declined; they received 3.3, 2.8, and 2.3 percent, respectively. Also, contrary to all the rhetoric about agriculture being a key priority area, its share in the Third Plan (6.7 percent) translated into a decline of 31 percent from its share in the Second Plan (9.7 percent).

The ambitious nature of the Third Plan becomes clearer when one realizes that the Second Plan (1970–74) called for only ₦ 3.2 billion, whereas the Third Plan called for a total investment of ₦ 30 billion. It shows the feelings of the government of the day: money was not to be a constraint. In fact, the Third Plan (FRN 1975a:48) stated that "there will be no savings and foreign exchange constraints during the Third Plan period and beyond." The drafting of the plan took place during the oil boom, when oil prices were increasing and demand was on the rise. Based on these conditions, it was possible to assume a position independent of foreign aid and external borrowing, and even become an aid donor (FRN 1975a:39, 364).

One of the objectives of the Third Plan was a continuation of the indigenization activities that were started under the Second Plan. Despite the optimism expressed during 1974 about the result of the indigenization exercise, by mid-1975 it became clear that several "legitimate" ways had been devised to circumvent the provisions of the 1972 decree. In fact, soon after he came to power, General Mohammed's government set up several panels to study Nigeria's economic and social problems. The Industrial Enterprise Panel Report admitted that, as of mid-1975, the implementation of the 1972 decree "fell short of expectation" (FRN 1976a:4). Apart from exemptions granted, a total of 950 firms were marked for indigenization: Schedule I (100 percent Nigerian participation) numbered 357 and Schedule II (40 percent Nigerian participation) numbered 593. After proper scrutiny, only a total of 314 enterprises of those firms had complied with the 1972 decree. Above all, no violator had been penalized. The Industrial Enterprises Panel Report (FRN 1976a) described the situation this way:

The main devices employed to circumvent the provisions of the Decree included fronting, application for naturalisation, extended use of the definition of Nigerian citizenship, interpretational problems of classification of enterprises, the gentle approach to implementation of the Decree and frequent amendments providing for exemptions on flimsy grounds. In almost all instances, the devices employed by the foreign owners could not have worked without the active support and connivance of some misguided Nigerian citizens.

To enforce compliance with the 1972 decree, July 15–17, 1976, was set aside for sealing up all the operations of defaulters and for freezing their accounts. Their merchandise was to be auctioned off on July 18. It was later decided that it was better to sell defaulting businesses as ongoing operations to Nigerians than to auction off their merchandise. The operation was carried out. On Monday, July 19, 1976, the businesses were reopened under the management of the foreign owners and the government-appointed co-managers.

After that 1976 operation, the Nigerian government of General Obasanjo promulgated the Nigerian Enterprises Promotion Decree on January 12, 1977 (*Daily Times* January 15, 1977; *Business Times* January 18, 1977). The new decree repealed all five previous versions of the 1972 decree. All business activities in Nigeria were covered by three schedules of the 1977 decree, and no exemptions were to be granted. Schedule I contained 40 economic activities reserved exclusively for Nigerians. Schedule II contained a list of 57 businesses that were required by the new decree to have 60 percent Nigerian equity participation rate (EPR). Schedule III embraced 38 business activities (not covered by Schedule I or Schedule II) that were to have at least a 40 percent EPR. The deadline for compliance was December 31, 1978, even though the new decree took effect June 29, 1976.

By not specifying how the equity shares were to be transferred to Nigerians, the 1972 decree only served the purpose of helping a few Nigerians with good business and political connections to acquire most of the shares of profitable foreign firms and become company directors while the majority of Nigerians were left empty-handed, thus increasing the already highly skewed distribution of wealth in Nigeria. The 1977 decree, therefore, sought to rectify the prevailing inequality in the distribution of income and wealth. It stipulated that no more than 5 percent equity of a foreign business or no more than N 50,000, whichever is greater, should be sold to any one Nigerian. Also, all Schedule II and Schedule III businesses were required to offer at least 10 percent of their shares for sale to their employees, and nonmanagement employees must purchase at least half of those shares.

The 1977 decree also restructured the Nigerian Enterprises Promotion Board (NEPB). It was to consist of 12 members, 5 of them appointed from outside the public sector in order to reduce the influence of the bureaucrats. The board also had its first full-time executive chairman. The NEPB was given additional authority to deal with defaulters and other indigenization offenders. Any person who acted as a front, or falsely pretended to be the owner or part owner of any business, or operated a business on behalf of a foreigner would be liable to a fine of N15,000 or imprisonment for 5 years (*New Nigerian* 1977).

Under the 1977 decree, nearly 1,200 foreign businesses were expected to sell their shares to Nigerians by the end of 1978. In 1977, when 12 of the well-known foreign companies offered over 82 million shares for sale, Nigerians proved that lack of capital was not a problem. The oil boom had created a new investor class among Nigerians. The demand for shares offered for sale through subscription exceeded the supply by as much as 4.3 to 1 in some cases. The number of securities listed on the Lagos Stock Exchange rose from 49 in 1970 to 107 in 1978, with industrials rising from 19 to 60 during that period (Lagos Stock Exchange 1979).

In sum, during 1975–79, the Nigerian regime focused attention on the return to civilian rule, the implementation of the Third Plan, and the enforcement of the 1977 Indigenisation Decree. In 1975, Nigeria's foreign reserves stood at $5.5 billion. By 1978, they had been depleted to $1.9 billion before the Obasanjo regime took austerity measures that raised their level to over $5 billion in 1979. During 1975–79, the economy remained dependent on oil exports as the main source of government revenues.

THE SHAGARI REGIME: OCTOBER 1979–DECEMBER 1983

As a result of the 1979 elections, Shehu Shagari was sworn in as the new head of state of Nigeria. He was able to gain the support of Azikiwe's NPP to achieve a working majority in both the federal Senate and the federal House of Representatives. For his Cabinet, 24 ministers were appointed. He also nominated 14 non-Cabinet ministers and 10 special advisers. The most controversial appointments made by the president were those of presidential liasion officers (PLOs) to the states. These PLOs were viewed, particularly in the non-NPN states, as NPN governors.

In May 1981, President Shagari granted a free pardon to Col. Odumegwu Ojukwu, the leader of Biafra during the civil war. It has been speculated that the underlying motive for this was the president's hope that Ojukwu would help shore up the ruling NPN after the breakup of the alliance between the NPN and the Eastern Region-based NPP. On the other hand, there are reports that Shagari had urged Ojukwu not to reenter politics so soon after his return from exile in the Ivory Coast, and that by joining the NPN, Ojukwu embarrassed Shagari (*TIME* 1983:59).

The Shagari government came to power during a period of rapid petroleum price increases that masked the decline of petroleum production. Average daily petroleum production was 2.3 million b/d in 1979. In 1980, it declined to 2.1 million b/d, to 1.43 million b/d by 1981, and to 1.2 million b/d in 1982 and 1983. Despite the decline in production, the value of petroleum exports increased from N-10.04 billion in 1979 to N-14.0 billion in 1980. In 1981, despite the sharp decline in production, total petroleum exports were valued at N-11.3 billion, and N-10.5 billion in 1982, before dipping to N-7.7 billion in 1983. During 1979–83, petroleum accounted for over 90 percent of all exports (IMF 1984:454–55).

After Nigeria had rapidly depleted its foreign reserves in the boom years 1976–78, the Obasanjo regime imposed restrictions on imports that increased foreign reserve holdings from N-1.9 billion in 1978 to N-5.5 billion in 1979, when Shagari took office. By 1980 reserve holdings had nearly doubled to N-10.2 billion (IMF 1984:452– 53). The more than doubling of reserve holdings partly reflected the effects of the austerity measures instituted by the Obasanjo regime, which were continued during the first year of civilian rule. Between 1981 and 1983, Nigeria's foreign reserves were depleted once more—by 1983, they had dropped to less than N-1 billion.

Petroleum revenues were important for paying for imports and for financing Nigeria's Fourth National Development Plan (Fourth Plan) for 1981–85. The outline of the Fourth Plan, as presented in 1981, emphasized the promotion of self-reliance and the involvement of the masses in the development process. The practical implications of greater self-reliance are not clearly specified, but one would expect it to include the substitution of Nigerian for foreign factors of production, products, services, techniques, and tastes. It proposed public investment of N-70.5 billion (N-14.1 billion per annum). The highest priority was placed on transportation, which received 17.2 percent. It was followed by trade and industry (including mining) with 16.3 percent. Education placed third with 10.7 percent, followed by general administration and defense with 8.8 percent. Agriculture received the fifth largest share of investment (7.9 percent), an increase of just over 1 percent above its reduced share in the Third Plan (6.7 percent). Electricity, health, water, town and country planning, and communication received 4.7, 4.3, 4.1, 3.8, and 2.8 percent, respectively.

The Fourth Plan was based on a projected petroleum production of 3.0 million b/d—which never materialized, because of the decline in world oil demand and more competition from non-OPEC oil suppliers, such as Britain, Norway, and Mexico. The result was a curtailment in plan activities and reduction—in many cases, stoppage—of several development projects.

The return of Nigeria from military rule to a democratically elected government was seen as a victory for democracy. However, the political history of Nigeria between October 1, 1979 (when President Shagari took the oath of office) and December 31, 1983 (when he was ousted in a military coup) leaves much to be desired. There was widespread corruption in the ruling party at the center, and by the various state governments. *South* (1984b:19–20) had this comment:

> Though the coup's cutting edge was directed at the NPN federal rulers, it was set against the brand of politics practiced by all parties of the Second Republic. For although the NPN dwarfed its rivals in corruption, mismanagement and election rigging, they all had, in the states they governed, demonstrated similar behaviour. Many Nigerians believe, on evidence from the 19 states, that the comparatively mild excesses of the other parties was limited not by virtue but by lack of opportunity.
>
> . . . Meanwhile high-level, wholesale looting of the impoverished national treasury continued unabated. In fact, the various task forces were designed to help the ruling party and its chief agents in their habitual plunder of the nation. Party chieftains or their agents were given licenses and government funds to import goods. They did so at highly inflated prices which enabled them to divert into private accounts abroad billions of dollars of scarce foreign exchange.
>
> When the public complained of hunger, Umaru Dikko, the strongman in the cabinet, retorted that things were not that bad since the people were not yet eating off rubbish heaps. The NPN also undertook to subvert the political system in order to entrench itself. To accomplish what was no less than an electoral coup in August, it bent to its will every organ of the state which was supposed to arbitrate the contest impartially.
>
> . . . The parliamentary democracy of the First Republic failed; and there is a strong feeling now that no misgovernment could be worse than that of Shagari's presidential democracy.

The *Wall Street Journal* (1984:17) described the hardships experienced by Nigerians under the Second Republic:

> The doubling of the price of rice after the reelection of Mr. Shagari illustrates the hardships of the average Nigerian as well as the government's corruption. President Shagari then set up a special task force, which was to relieve the problem by importing and distributing rice.
>
> The group brought in the rice, but it hoarded much of it in warehouses and sold it at grossly inflated prices. U.S. diplomats say rice was reaching Nigerian docks at $480 a ton, but it was being sold on the street at more than five times that amount.
>
> "Someone was making a lot of money somewhere," one U.S. diplomat says.
>
> After the coup, military authorities assert that they found bags of rice stashed away in the home of the head of the presidential task force,

the minister of transport and aviation, Umaru Dikko . . . who is thought to have fled to Europe.

Furthermore, there was an acute shortage of parts and components for industrial production. According to the News Agency of Nigeria, by the end of August 1983, 49 out of 73 industries that were operating members of the Manufacturers Association of Nigeria (MAN) had closed down. And since April 1983, 107 factories had temporarily closed for 4 to 8 weeks, and reduced production to three days a week resulting in increased unemployment (*West Africa* 1983c:2538).

On December 29, 1983, President Shagari presented his 1984 budget. Even though he had earlier declared a national austerity, the budget projected a deficit of U.S. $1.65 billion, which was to be financed by domestic and foreign debt. The existing national debt already amounted to some U.S. $15 billion (*South* 1984b:20). Out of projected foreign exchange earnings of U.S.$6 billion, U.S.$2.25 billion would be used to service existing debt, leaving only U.S.$3.75 billion for the importation of essential commodities. Perhaps the people would have understood the economic situation in Nigeria were it not for the ostentatious living of the ruling political class, as acknowledged by Shagari himself when he asked his legislators, "How can you convince . . . . your constituencies that there is austerity when cars of various brands are seen parked in your houses?" (*West Africa* 1983b:2480).

Hence, the end of the Shagari regime in Nigeria was welcome news to most Nigerians. It was expected. In fact, the retired army chief of staff, Gen. T. Y. Danjuma, a member of the military government that had handed over power to Shagari, was quoted as saying: "After the 'landslide' [victory of the NPN in August] I knew there will be a 'gunslide'." (*New African* 1984:55).

THE BUHARI REGIME: JANUARY 1984–AUGUST 1985

By 1984, the economy of Nigeria was in shambles. Therefore, the overthrow of the civilian government was welcome news to Nigerians. That military coup brought Maj. Gen. Muhammadu Buhari to power. Not only did the regime inherit declining revenues based on fluctuating petroleum sales, but it was faced with mounting foreign and domestic debts.

On coming to power, the Buhari regime declared war on corruption and indiscipline throughout Nigeria by means of a much publicized program christened W.A.I (War Against Indiscipline) (*West Africa* 1984:793), and began to address the accountability of politicians of the last civilian government. In its headline, *TIME* (1984c:52) said of the Buhari regime: "Rooting Out Corruption: Offering No Apologies, a New Leader Presses

His War Against Indiscipline." Since coming into office, the Buhari regime has recovered large sums of money from former politicians.

Furthermore, the administration maintained a firm stand on the IMF loan. During Nigeria's negotiation for a loan in 1984, some of the IMF's requirements were devaluation of the naira, cutting petroleum subsidies, and trade liberalization. The consensus in Nigeria was that these conditions were detrimental to the already ailing economy. Labor leaders warned that the IMF conditions would lead to higher prices for many consumer goods and services, such as gasoline, electricity, and transport, as well as to the closing of more factories and greater foreign domination of the economy (*South* 1984a:24). The feeling was that the poorest people were likely to be hit hardest by the IMF requirements.

In his interview with the editor of *West Africa* (1984c:424–27), Major General Buhari challenged the logic of the IMF requirements:

> The way the IMF sees it, if we devalue our exports would be cheap, imports would be dearer. If so, the effect on Nigeria is irrelevant because we hardly export anything other than oil which is dollars and which is subject to currency fluctuations, so devaluation doesn't make sense, because our industries hardly satisfy our needs up to 50 percent. So we are not exporting anything other than oil; finished goods are second, so that argument does not hold. On the cost of imports, we need cheap imports, because our essential raw materials for our industry are mostly imported from the United States and Europe so we don't want to make it expensive. If we make it expensive our end product would be more expensive, and the inflation will go up again, so the argument against devaluation in Nigeria is real, and we hope the IMF will see it that way.

The regime felt that by refusing the IMF conditions, it had avoided mortgaging the welfare of future generations (*West Africa* 1984g:815–16; 1984k:1817).

As far as the development strategy of the Buhari regime was concerned, the fact that the regime lasted for only 20 months makes an evaluation of its policies more difficult. However, since it was the first government to come to power because the economy of Nigeria was in disarray, an evaluation of its success in the minds of most Nigerians would be affected by the fact that unemployment increased dramatically while the regime was in power, life became harder for most people, shortages of food and essential commodities persisted, and inflation continued to rise.

On August 27, 1985, after struggling for 20 months to cure the country's economic problems, General Buhari's government was overthrown in a military coup. Gen. Ibrahim Babangida was named as Nigeria's new president and military commander.

SUMMARY

Between 1970 and 1985, Nigeria was governed by five military governments and the civilian government of Shehu Shagari. During the same period, the strategy of indigenizing the Nigerian economy was carried out. The two Indigenisation Decrees were attempts to increase Nigerian participation in the ownership and control of essential business activities. The result had been the employment of Nigerians in every firm operating in the country. No doubt that indigenization was perceived as a vital part of economic development, and was expected to produce a self-reliant and economically independent country. Unfortunately, that had proved to be elusive. Instead, Nigeria became more dependent economically because of its international trade and oil export orientation.

The oil boom and the indigenization policy greatly benefited the middle class, while living conditions became more difficult for the lower class (the majority of the population). The oil boom of the 1970s gave Nigerian leaders and planners a false sense of hope. Foreign exchange ceased to be a bottleneck during the period because of massive inflows of new capital. In fact, the unexpected oil boom gave the regimes of the 1970s and the early 1980s the idea that Nigeria could be transformed into a modern country overnight.

Indeed, the appearance of development was everywhere during the 1970s. For example, road building in Nigeria experienced tremendous growth between 1970 and 1980. The total road network increased from 66,074 kilometers in 1960 to 95,374 kilometers in 1972, an increase of 44 percent (FRN 1974a; 1975a). Since then, major projects have been undertaken to develop a national integrated highway system. It has been estimated that between April 1975 and December 1978, the combined length of roads and bridges under construction increased from 4,800 kilometers to 14,500 kilometers (FRN 1979a:xix). The importance of good roads in Nigeria cannot be overemphasized when one realizes that between 1978 and 1980, the motor vehicles in use increased from 449,424 to 633,268, an increase of almost 41 percent (*Africa South of the Sahara* 1983:651). Also, road transport is a major tool of interstate commerce. During the 1960s, the Nigerian Railway Corporation operated two lines: Lagos to Kano (1,100 km.) and Port Harcourt to Maiduguri in the north (1,435 km.). By the mid-1970s, the length had grown to around 3,500 kilometers. The growth of the railroad would have been more dramatic were it not for the advantages that road transport enjoyed over it—road transport is relatively cheaper and provides door-to-door service.

Education in Nigeria is regarded as an instrument of development. This popular view was confirmed by Chief Obafemi Awolowo, who in

1955 introduced free primary education into the Western Region. He wrote in his memoirs:

> the provision of education and health in a developing country such as Nigeria is as much an instrument of economic development as the provision of roads, water supply, electricity, and the like. To educate the children and enlighten the illiterate adults is to lay a solid foundation not only for future social and economic progress but also for political stability. A truly educated citizenry is, in my view, one of the most powerful deterrents to dictatorship, oligarchy and feudal autocracy. (Awolowo 1960:268)

According to the Third Plan (FRN 1975a:238), grade school enrollment was expected to grow from 2.9 million in 1960 and 3.9 million in 1971 to 11.5 million during 1975–80 as a result of the introduction of free universal primary education. The increased enrollment will no doubt create additional demands on the government, in the 1980s and beyond, for more secondary schools, more universities, more jobs. As we have seen in this chapter, however, most of Nigeria's investment had gone into physical infrastructure development rather than development of productive sectors. Physical infrastructure development means little to Nigerians if they cannot assume positions of industrial leadership.

Despite the appearance of development, the oil boom of the 1970s did not result in economic and political autonomy for Nigeria. Rather, the economy became more dependent on oil revenues. Billions of naira were spent on physical infrastructure. Still, Nigeria had little to show for all the money because of waste and corruption on a large scale.

Electricity generation and distribution in Nigeria are controlled by the National Electric Power Authority, a public corporation that came into existence in 1972 as a result of the merger of the former Electricity Corporation of Nigeria and the Niger Dams Authority. Even though production of electricity in Nigeria started in 1898, it was not until after 1960, especially after the completion of the Kainji Dam, that generating capacity experienced substantial growth (IBRD 1955; Mabogunje 1973). Table 7.8 shows the generating capacity between 1963 and 1982. According to the Third Plan (FRN 1975a:175), consumption of electricity had been growing at more than 20 percent per annum, whereas the annual growth rate of generating capacity between 1960 and 1980 was 15.8 percent. The result was irregular and unreliable electricity supply and regular blackouts throughout the country.

Communication systems in the 1970s were extremely inefficient and poor. Telephone and telegraph lines were not transmitting most of the time, and when they functioned, they did so unsatisfactorily. The Post and Telecommunications Department (of the federal government) had

TABLE 7.8. Electricity Generating Capacity, Selected Years: 1963–82 (megawatts)

Year	Installed Generating Capacity
1963	217.0
1966	368.0
1970	804.7
1971	804.7
1972	786.7
1973	793.7
1974	793.7
1975	810.7
1976	912.2
1977	1,064.5
1978	1,729.1
1979	1,822.7
1980	2,230.5
1981	2,430.0
1982	2,904.1

Source: Central Bank of Nigeria, *Annual Report & Statement of Accounts,* various volumes.

the monopoly over telephone and telegraph. In 1960, Nigeria had a total of 18,024 telephones. With the oil boom, and with a population close to 80 million people, in 1977 there were only 55,000 telephones. The lack of telephones meant increased messenger services and increased costs of vehicle maintenance, millions of man-hours wasted on the highway, plus undue annoyance. Nigeria's telephone density was, and remains, one of the lowest in the world—a clear roadblock to its development (*Business Times* 1978:24).

Despite the oil boom of the 1970s, sewerage systems were nonexistent; hospitals lacked essential medical facilities; most of the store shelves were empty; and unemployment increased dramatically. As Zartman and Schatz (1983:15) have rightly pointed out, the government sector expansion neglected to ask whether substantial increases in government consumption and military expenditures were what Nigerians most wanted or Nigerian society most needed.

The decline of agriculture betrayed the urban bias of the development plans. The portion of the labor force in agriculture decreased from 71 percent in 1960 to 54 percent in 1980 (World Bank 1983b:II, 69). Also, the average rural income, which was 38 percent of urban income in 1960, fell to 24 percent by 1977 (Zartman and Schatz 1983:15).

Political changes have accompanied both social and economic changes resulting from the oil boom. Political power in Nigeria has become more centralized. This is evident in the control over national revenues, industry, produce marketing, and college admissions. As Prof. Claude Ake (1981:1163) argued in his presidential address to the Nigerian Political Science Association:

> As things stand now, the Nigerian state appears to intervene everywhere and to own virtually everything including access to status and wealth. Inevitably a desperate struggle to win control of state power ensues since this control means for all practical purposes being all powerful and owning everything. Politics becomes warfare, a matter of life and death.

Nigeria's emphasis on trade and capital accumulation resulted in growth without development. Autonomy, as expressed in the various plans, was focused on Nigerian ownership of enterprises while neglecting the main meaning of autonomy in an economic and political sense, which would emphasize the ability of a nation to stand on its own: to be self-sufficient, and able to exist independently, without outside dominance.

Chapter 8 provides a critical analysis of national development in Nigeria from 1960 to 1985, with particular emphasis on the oil boom of the 1970s. It also evaluates the internal and external factors that have affected national development in Nigeria.

NOTES

1. The planners sought to explain away this latter bias in the First Progress Report on the Second Plan by saying, "These distortions arise mainly from the fact that low priority projects in the administrative sector are relatively easier to prepare and execute than projects in the productive sectors" (FRN n.d.:39).

2. As Akinsanya (1983b:157) points out, Nigeria voted for several U.N. General Assembly resolutions stating that a state which exercises the right to

nationalize, expropriate, or transfer ownership of foreign property shall pay the deprived owner "appropriate" or "possible" compensaton.

3. While critics view Gowon's success in foreign policy as a result of his ability to divert attention away from domestic problems, it is nonetheless worth mentioning that he was instrumental in the signing of the Treaty of the Economic Community of West African States by its 16 member countries.

4. Other actions taken by the Mohammed government included the postponement until February 1977 of the Second World Black Festival of Arts and Culture (FESTAC), which had earlier been unrealistically scheduled for the end of 1975. (The FESTAC eventually cost the federal government N-140 million.) Also, the cement scandal, in which a few distributors became multimillionaires by overordering millions of tons of cement, was resolved. Until Mohammed's action, the cement importation scandal meant that over 400 ships were at anchor at Lagos, running up demurrage charges until it was their turn to offload their commodities. He also canceled a plan to build a huge State House at Victoria Island, at a cost of N-20 million; and he stopped plans to build new residences for several state governors (Kirk-Greene and Rimmer 1981:13). Because of his actions against corruption, he was popular throughout the country. The concept of "Ramatism" (his middle name was Ramat) was introduced to stand for a leader of the common man and an opponent of corruption.

8 A Critical Analysis of Nigeria's Costly Progress

Everyone knows an underdeveloped country when he sees one. It is a country characterized by poverty, with beggars in the cities, and villagers eking out a bare existence in the rural areas. It is a country . . . with inadequate sources of power and light. It usually has insufficient roads and railroads, insufficient government services, poor communications. It has few hospitals and few institutions of higher learning. Most of its people cannot read or write. It may have isolated islands of wealth, with a few persons living in luxury. Its banking system is poor; small loans have to be obtained through moneylenders. . . . Its exports. . . usually consist almost entirely of raw materials, ores or fruits or some staple product, with possibly a small admixture of luxury handicrafts. Often the extraction or cultivation of these raw material exports is in the hands of foreign companies.

(*One Hundred Countries—One and One Quarter Billion People:*14 [Washington, D.C.: Committee for International Economic Growth, 1960], as quoted in Morgan 1975:37).

INTRODUCTION

To what extent is the Nigerian development experience explained by internal political, economic, and social factors, such as Nigerian nationalism, the political dominance of the military, the Biafran civil war and its aftermath, and military and civilian government corruption? And to what extent is its development pattern the result of a structural condition of underdevelopment created by its colonial past and continued by its export petroleum economy?

Can Nigeria's experience be explained comprehensively by its decision to pursue unbalanced growth, or must the influence of developed economies be incorporated to explain the experience? These issues are the focus of this chapter.

THE DEVELOPMENT CRISIS IN NIGERIA

Since Nigeria attained political independence in 1960, its economy has undergone dramatic and highly uneven change. The growth of the petroleum economy since 1964, even though its role in government revenues and capital accumulation did not become evident until after the oil price increases of 1971, was the driving force of Nigerian economic change and the source of its uneven pattern of development.

Table 8.1 presents the average annual growth rate of Nigeria's main economic sectors for 1960–70 and 1970–81. It shows that the greatest growth between 1960 and 1970 occurred in the mining sector, where petroleum is the dominant commodity. In 1970–81, mining grew at an average annual growth rate of only 2.5 percent, a sharp decline from an annual growth rate of 31.7 percent in 1960–70. Petroleum production in-

TABLE 8.1.   Sectoral Average Annual Growth Rate, 1960–81

Sector	*1960–70*	*1971–77*	*1970–81*
Agriculture	1.3	-1.5	-0.4
Mining	31.7	8.1	2.5
Manufacturing	12.8	13.4	12.4
Construction	9.0	24.7	13.1
Electricity, gas, water	16.7	18.2	16.3
Transport, communication	4.0	16.5	8.3
Public Administration, defense	4.8	24.6	5.7
Imports of goods and services	4.0	24.7	15.9
Exports of goods and services	12.9	5.6	0.8
GDP, at factor cost	4.3	6.2	4.5
GDP, at market prices	4.4	6.0	4.5

Source: World Bank 1980:150–51; 1983b:134–35.

creased from a daily average of 0.418 million b/d in 1966 to 2.256 million b/d in 1974. After that, daily production fluctuated greatly. In April 1982, the average daily production dropped to 0.9 million b/d; the year ended with an average of 1.289 million b/d, then decreased to 1.235 million b/d in 1983 (FRN 1984c). Exports of goods and services followed a similar pattern, slumping from an average annual growth rate of 12.9 percent during 1960–70 to 0.8 percent during 1970–81.

Agriculture, the economic activity in which the majority of Nigerians are engaged, showed the lowest average annual growth rate during 1960–70 (a mere 1.3 percent), declined to -1.5 percent during 1971–77, and ended the period 1970–81 with an average annual decline of 0.4 percent. The decline of agriculture has played a pivotal role in Nigeria's recent economic life. Before the oil boom of the 1970s, Nigeria relied on agriculture exports (cocoa, cotton, palm oil, palm kernels, rubber, and timber) for 65 to 75 percent of its foreign exchange. By 1970, their value had dropped to 44 percent of total exports. In 1978–79, they accounted for only 6 percent of the total (*Africa South of the Sahara* 1984:639). In 1982, Nigeria relied on oil for 90 percent of its foreign exchange. Until 1973, Nigeria was the second largest cocoa producer in the world (after Ghana); it now produces about 11 percent of the world total. Similar declines have become evident in the other agricultural commodities, such as peanuts, peanut oil, palm kernels, palm oil, rubber, raw cotton, timber, and logs.

The consequence of agriculture decline is great for Nigeria. Agriculture provides employment for over half of the nation's work force. Its decline since 1960 has been followed by a substantial decline in the nation's labor force participation rate. In 1960, the proportion of the labor in agriculture was 71 percent. By 1980, it had declined to 54 percent. During the same period, the proportion of the labor force in industry increased from 10 percent to 19 percent. However, Nigeria's total labor force participation rate declined from 42.2 percent in 1960 to 36.4 percent in 1980 (World Bank 1983b:II, 69). In addition to providing food for the population, and raw materials for the industrial sector, agriculture has traditionally generated most of the revenues and foreign exchange from exports. The decline in this sector not only has adversely affected the exports but also has led to increases in imports, which in turn affects Nigeria's foreign exchange reserves. In 1970, food constituted 7.6 percent of all imports. By 1974, had risen to 8.9 percent, and by 1978, it had increased to 12.4 percent of all imports.

For a decade, Nigeria's oil masked the structural problems of the economy. Its foreign exchange reserves increased rapidly from $0.355 billion at the end of 1972 to $5.602 billion by 1974. The main contributor to the increase was petroleum revenues. It was, therefore, not surprising that when petroleum production dropped in 1978, foreign exchange reserves fell to $1.887 billion—less than enough to cover two months' import bills. With higher petroleum prices in 1979 and 1980, foreign exchange reserves recovered, but only temporarily. In 1980, reserves rose to $10.24 billion. Following the slump in the oil market during 1981 and 1982, the foreign exchange reserves dropped drastically. By the end of 1981, they were down to $3.9 billion. The decline continued—by 1982, reserves were down to $1.61 billion, and by 1983, were $0.99 billion (IMF 1984). The sudden fluctuations and steep declines in petroleum revenues have destabilized Nigerian development.

Many sectors of the Nigerian economy depend upon imports in order to produce goods, and thus growth in imports is to be expected. However, this means that the economy is structurally organized to exacerbate its foreign exchange problem; export earnings, determined by world oil market conditions, will vary greatly, while foreign exchange needs will continue to grow. Left on its own, the Nigerian economy will drift from crisis to crisis. Yet the national government is unable to correct the imbalance through its revenues because its primary source of funds is the faltering petroleum economy. Federally collected revenues in 1983 were ₦ 8.6 billion, a drop of 21.8 percent from ₦ 10.9 billion in 1982. The decline in government revenues has compelled the federal and state governments to utilize deficit financing by borrowing heavily from the banking sysetm. According to General Buhari's May 1984 Budget Speech, in 1983 alone the credit extended to the public sector was ₦ 5.293 billion. Also, there was a backlog of unpaid wages and salaries of workers, and substantial amounts owed by federal and state governments to contractors and suppliers (FRN 1984c). Public internal debt increased from ₦ 4.636 billion in 1977 to ₦ 15.011 billion in 1982, and to ₦ 22.221 billion in 1983. External debt rose from ₦ 1.131 billion in 1977 to ₦ 5.341 billion in 1982 (West Africa 1984e:719). As of March 31, 1984, external debt stood at ₦ 8.3 billion, out of which the federal government accounted for ₦ 5.32 billion and the states ₦ 2.98 billion (FRN 1984c).

Former Minister of Finance Dr. Onaolapo Soleye, in his statement on the 1984 budget (FRN 1984c:13), indicated that Nigeria was engaged in negotiations with the IMF for a balance of payments support loan of ₦ 1.9–2.46 billion. Even though progress was reportedly being made, the unresolved issues were devaluation of the naira, trade liberalization, and removal of petroleum subsidies. Also, the proposed World Bank-structured adjustment loan of ₦ 224 million per annum for Nigeria in 1984 and 1985 was linked with the IMF terms (FRN 1984c:13). In a ten-year period, Nigeria has gone from high growth with extraordinary domestic capital reserves to a near bankrupt economy trying to borrow its way out of collapse.

What went wrong? To answer this question, this chapter examines, first, the role of internal political, social, and economic factors and their influence on Nigerian strategies of national development. Specifically, it takes into consideration the following factors that shaped national strategies: the ideology of nationalism; the development theories and development planning apparatus utilized by the regimes to implement their plans; and certain external factors.

THE ROLE OF NATIONALISM IN THE NIGERIAN ECONOMY

Anticolonial nationalist movements played a major role in securing Nigeria's political independence from Britain in 1960. These movements were predicated upon the idea of national identity, which was denied by colonial rule. Nationalism for the anticolonialist leaders and for the mass of people that followed them was not mere symbol or ploy. As Sathyamurthy (1983:37) argues:

Nationalism in a general perspective is not a disaggregated incident in the life of a society; it is a facet of a continuous process of change. A new nation does not emerge from nothing or spontaneously. At the heart of the dialectic of decolonisation is the persistence of the indigenous community; traditional structures and values which have weathered the colonial storm are crucial to the unfolding of a new nation. Nationalism is balanced on a fulcrum which is signified by a process of survival and reconstruction containing a variety of alternative potentialities and manifested at several social levels.

Brenton (1964) argued that nationalism leads societies to invest resources in nationality or ethnicity, and that it encourages demands for changes in the international or interethnic distribution of property and wealth of a country. Harry G. Johnson (1967a:4) argued that

[N]ationalism will tend to favor investment in national manufacturing, since manufacturing jobs and ownership are preferred by the middle class; its collective nature will appeal to socialists; and its emergence will be correlated with the rise of new middle classes.

These arguments suggest that nationalist economic policies tend to emphasize activities such as manufacturing, particularly certain symbolic industries, such as steel, that suggest the nation is modern and moving toward a condition of self-sufficiency. Moreover, nationalist economic policy tends to support extensive state control over, and public ownership of, economic enterprises. Also, according to Daly and Globerman (1976), nationalist economic policies tend to redistribute material income from the lower class toward the middle class, particularly the educated middle class.

The Indigenisation Decrees of 1972 and 1977 exemplify nationalist policy. The purpose of the 1972 decree was to redistribute equity shares from foreigners to Nigerians. The resultant wealth concentration was not envisioned. The 1977 decree was an attempt to reduce the concentration of share equity ownership in a few hands. It is doubtful, however, whether that goal was ever achieved. Most of the shares of foreign investors were sold by private placement or offer for sale (NEPB 1979b). That gave the rich and the middle class who had political and business connections the opportunity to benefit most from the exercise. Participation by the general population was limited because of the low rate of offer for subscription.

The oil industry was affected by the indigenisation exercise. In 1971, the state-owned Nigerian National Oil Corporation (NNOC) was established to own equity shares in the main oil companies and to carry out exploration jointly with the foreign companies. In 1977, the NNOC was merged with the Federal Ministry of Petroleum Resources to form the Nigerian National Petroleum Corporation (NNPC). In July 1979, the NNPC increased its equity shares in oil operations from 55 percent to 60 percent. On July 31, 1979, in retaliation for BP's oil arrangement that led to Nigerian oil being exported to South Africa, BP's interests in Nigeria were nationalized, effective August 1, 1979 (Aluko 1981). Shell-BP's concessions amounted to more than 80 percent of all the oil concessions granted in Nigeria, and were yielding an average of 60 percent of the total oil production in Nigeria.

No doubt the indigenization exercises of 1972 and 1977 were designed to achieve self-sufficiency and to alter the equity ownership structure in the economy. They were also meant to increase Nigerian participation in the management of companies, especially the number of Nigerians employed in middle and top management. There were signs, however, that the stringent requirements of the Indigenisation Decree may be selectively relaxed in the near future. For instance, in the 1984 Budget Speech delivered by General Buhari, he announced that the federal government was considering "a proposal to amend the Nigerian Enterprises Promotion Decree to enable non-Nigerians to own up to 80% of large farm projects" (FRN 1984c:14).

Whatever changes Nigeria decides to make in its indigenization policy, the goal of becoming self-sufficient and economically independent via such a policy is likely to remain elusive. Indeed, the Nigerian economy of the 1980s is more dependent than ever. Its economic health is at the mercy of petroleum export, which provides over 90 percent of its foreign exchange earnings. The nation has found that with lower oil sales, manufacturing generally is adversely affected, and importation of food and other essential commodities is extremely costly. Whatever indigenization was supposed to accomplish, it did not seriously affect the dependency of the society. Nigeria is an independent nation in a legal sense, but its social and economic structure remains remarkably similar to that of its colonial past. According to Adeoye Akinsanya (1983b:179):

> the indigenization of private foreign investments under the Nigerian Enterprises Promotion Decrees of 1972 and 1977 has not really enhanced national control over the economy nor does the Federal Government's majority equity interests in foreign owned oil producing and insurance companies and banks necessarily confer effective control of the Government over the operations of the joint ventures.

In describing the same decrees, another Nigerian scholar argued:

> The Nigerian Business Enterprises Decrees . . . when taken in a holistic framework amount to no more than the indigenisation of the exploitation of the Nigerian nation and resources by the affluent members of the national bourgeoisie. (Oyebode 1977:15)

In assessing the role of nationalism in Nigerian development, one needs to distinguish between the results of political nationalism and economic nationalism. There can be little doubt that political nationalism in Nigeria was instrumental in the attainment of political independence in 1960. Economic nationalism, however, was less successful. Rather than producing economic autonomy, it may have been an impediment to national autonomy insofar as it fostered the view that autonomy can be achieved by the simple act of Nigerians acquiring ownership and management of private foreign companies operating in Nigeria. In this way, economic nationalism prevented Nigerian planners and leaders from focusing on the structural problems of the petroleum economy. Being patriotic is one thing; being responsive to economic reality is another. But while economic nationalism deflected leadership attention from the real causes of the dependent structure of the Nigerian economy, it certainly was not the principal obstacle to achieving autonomous development.

DEVELOPMENT PLANNING IN NIGERIA

Development planning in Nigeria, as in most other developing countries, is based on the neoclassical paradigm. This framework has been criticized as lacking historical specificity, in that it evades the discussion of the specific historical position that underdeveloped countries occupy within the polarized structure of the world capitalist economy (Mabogunje 1981). It has also been criticized for its failure to consider the effects of capital accumulation on the spatial structure of nations that are involved in the process of development (Mabogunje 1981). One of the effects of capital accumulation and the penetration of foreign capital is the excessive concentration of industrial activities in a few urban centers. The trend was started during the colonial era and continued after independence. This lack of appreciation for the fact that development requires a strategy of spatial reorganization has contributed to the distortions that planning based on neoclassical models has produced in Nigeria. The Nigerian planning apparatus sought to incorporate elements of social welfare, but their provision usually favored the urban dwellers. Inasmuch as most of the services subsidized by government revenues are located in the urban areas, the beneficiaries are the urban dwellers, whom Adebayo Adedeji (1973:14–15) called the "vocal minority." The welfare of the rural people, "the silent productive majority," tends to be ignored. Thus, the apparatus' effort to address social welfare goals served to exacerbate still further the tendency toward spatial overconcentration.

The planning strategy for industrialization in Nigeria involved the replacement of imported goods with machinery and equipment that produced the goods that previously were imported. The strategy of import substitution was adopted because it was felt that by starting with this type of industrialization, a country would eventually be led by backward linkage effects to develop domestic production of intermediate and basic industrial components. The import substitution industries received a high degree of protection from the state. This strategy resulted in raising the price of industrial products vis-à-vis that of agricultural commodities. Also, the artificially high exchange rate penalized agriculture by reducing the receipts in domestic currency from a given volume of agricultural exports (Mabogunje 1981). This has led to the decline of agriculture. Rather than decreasing the volume of imports, import substitution industrialization tended to increase it in Nigeria.

In addition, the planning strategy of import substitution industrialization led to the redistribution of population caused by the migration of rural dwellers seeking employment to the urban centers, particularly to Lagos. It was estimated that the population of Lagos grew from approximately 1.0 million in 1965 to about 2.5 million in 1975, an annual growth rate of 10.2 percent during 1965–75 (Mabogunje 1981:179). But because the machinery utilized for import substitution manufacturing tends to be capital-intensive rather than labor-intensive, most of the newcomers to the urban centers remain unemployed. Rather than reducing poverty, the strategy has exacerbated the situation. The strategy created an urbanization in which economic activities are growing too slowly to ensure employment for the rapidly increasing urban population. Breese (1966:5, 99) called such city growth "subsistence urbanization." As Aluko (1973:213–14) put it:

> While increasing migration caused mass unemployment in the urban areas, there is a growing shortage of labour and diminishing productivity in the rural areas, particularly in the agricultural sector which is regarded as the main occupation of rural people.

The colonial era no doubt started the lopsided rural-urban migration. However, it was the nationalist pursuit of the industrialization policies and the operation of an urban-biased income structure that increased it (Igbozurike 1976). Also, the tripling of oil prices in 1973 caused a wage explosion and a massive urban influx of rural workers (Morgan 1984). Igbozurike (1977) has estimated that well over 80 percent of Nigeria's urban population is of rural origin.

Furthermore, the backward linkages anticipated from import-substitution industrialization have not materialized because of its enclave nature. The consequence of this is that even though there may be prosperity in the urban areas, particularly in the capital city, it does not necessarily produce vigorous economic activity at the periphery of the country. The majority of Nigerians rely on agricultural incomes, which are declining. Therefore, the majority of the population is becoming worse off, and the rural–urban income gap is widening. Also, the oil industry, which had generated tremendous revenues for Nigeria during the 1970s, is capital-intensive and thus generates relatively few jobs.

Nigeria has utilized the single-sector project planning models since independence. This neoclassical planning approach focuses state attention on levels of production, consumption, and growth in individual sectors. In it, potential growth in isolated sectors is assessed on its own merit. This assessment forms the basis upon which industrial projects are planned. The popularity of the single-sector project model is due, to a significant degree, to the lack of the adequate, timely, and reliable statistical data needed for a complete multisector planning. The single-sector project model often suffers from lack of internal consistency and even of overall feasibility. Rather than a well-coordinated program of action, the Nigerian national plans became a mere collection of assorted development projects with no apparent interconnections and full of inner contradictions. In the 1970s, during the oil boom era of planning with unlimited capital, matters worsened as the single-sector approach exacerbated the structural dependence of the Nigerian economy by giving the most consideration to those sectors supporting trade and capital attraction.

The first National Development Plan was intended to run from 1962 to 1968, and envisaged a total public investment of N-1.584 billion, an annual target of N-264 million. The Second Plan (1970–74) proposed greater public investment spending—originally N-2 billion, later revised to N-3.272 billion (N-818 million per annum). Less than 20 percent was expected to come from external sources. Public investment spending in the Third Plan (1975–80) was even greater. It grew enormously to N-43.3 billion (N-8.66 billion per annum. As a result of the rise of oil revenues, Nigerian planners and leaders thought that the nation could become independent of external borrowing and foreign aid. The surge in oil revenues in 1974 also persuaded the planners that savings and foreign exchange had ceased to be constraints on Nigeria's development. The outline of the Fourth Plan (1981–85), presented in 1981, proposed public investment of N-70.5 billion (N-17.6 billion per annum). This plan was based on a projected petroleum production of 3.0 million b/d. But the sudden decline in oil revenues necessitated the reduction or stoppage of several projects proposed under the plan.

TABLE 8.2. Sectoral Distribution of Public Sector Capital Investment, 1962–85 (percent)

Sector	1962–68	1970–74	1975–80	1981–85
Agriculture	13.6	9.7	6.7	7.9
Trade & industry (including mining)	13.4	7.3	26.0	16.3
Transport	21.3	21.3	22.2	17.2
General admin. (incl. defense)	7.1	22.7	13.6	8.8
Electricity	15.1	5.0	3.3	4.7
Town & country planning	6.2	2.2	9.2	3.8
Water	3.6	5.8	2.8	4.1
Communication	4.4	2.4	4.1	2.8
Education	10.3	11.0	7.5	10.7
Health	2.5	5.0	2.3	4.3
Labor, social welfare, sports	0.7	1.3	0.4	0.3
Information	0.5	0.9	0.12	0.8
Cooperative & comm. devt.	0.6	—	0.6	—
Judicial	0.1	—	—	—
Financial oblig.	0.6	2.1	—	—

Sources: Federation of Nigeria 1961, 1970, 1975, Africa South of the Sahara 1983.

TABLE 8.3.   Average Sectoral Distribution of Public Sector Capital Investment, 1962–85 (percent)

Sector	Avg.	Rank
Transport	20.50	1
Trade & industry (incl. mining)	15.75	2
General admin., defense, security	13.05	3
Education	9.88	4
Agriculture	9.48	5
Electricity	7.03	6
Town & country planning (incl. housing)	5.35	7
Water (other than irrigation)	4.08	8
Health	3.53	9
Communication	3.43	10
Labor, social welfare, sports	0.68	11
Financial obligations	0.68	11
Information	0.58	13

Source: Olayiwola 1985:367.

As Table 8.2 indicates, public sector capital investment was concentrated in the petroleum sector and the infrastructure sectors necessary to support the growth of petroleum. Table 8.3 shows that the transport, petroleum, and defense sectors received by far the highest priority throughout the four development plans (20.5 percent; 15.75 percent and 13.05 percent respectively). A graphical representation of the pattern of public expenditure in each sector during the four planning periods is presented in Figure 8.1.

As Mabogunje (1981:67–68) points out, development involves

intense concentration of activities requiring a high degree of synchronization and sequential ordering over a short period of time during which the whole social fabric of a country is transformed and its spatial structure reorganized. . . . Growth, on the other hand, is a process, involving different rates of incremental change during which the possibilities of a given organization are fully exploited . . . any organization has the capac-

FIGURE 8.1   Average Sectoral Distribution of Public Sector Capital Expenditures, Four National Development Plans
(percent)

Sectors	Percent	
Agric.	9.48	****************************
Trade & Ind.	15.75	**
Electricity	7.03	*********************
Transport	20.50	***
Commun.	3.43	***********
Water	4.08	***************
Education	9.88	*************************
Health	3.53	************
Town & Ctry.	5.35	******************
Labor	0.68	**
Info.	0.58	*
Gen & Defen.	13.05	********************************
Fin. Oblig.	0.68	**

Source: Provided by author.

ity for growth but only up to a point. To go beyond that point involves a concern with the development of the organization. This entails structural transformation to make the system better able to respond to the new demand for a high level of efficiency and equity.

The development planning undertaken by Nigeria was consistent with her choice of development strategies. Despite the ample and available capital flowing to Nigeria to support her chosen neoclassical development path, Nigerian development plans failed to effectively address the structural problem of an inherited colonial economy dependent upon foreign markets for earnings and imported technology and food. Insofar as the Nigerian leadership devised these plans, implemented them, witnessed their failures, and yet continued to plan according to the same principles, the crisis in which Nigeria finds herself appears to be at least in part the result of the leadership's unwillingnes or inability to recognize the serious flaws in the neoclassical paradigm. But again, this failure can only be judged to have added to Nigeria's problem. The planning apparatus did not in itself cause dependent development; rather, it substantially reinforced such a condition.

OTHER INTERNAL FACTORS AFFECTING NIGERIA'S DEVELOPMENT

There are several internal factors that affected, and continue to affect, Nigeria's efforts toward growth with development. Just six years after the nation gained political independence, the first military coup took place, not because of mismanagement of the economy, but because of political chaos. The main objective was the stabilization of the political system of the country. The second coup, which brought General Gowon to power, had the same objective. In fact, the Gowon regime had its hands full with the Nigerian civil war. Despite the oil revenues, which provided the funds for reconstruction, the civil war proved to be a big drain not only on the country's physical and economic resources but on its human resources as well. Resources that could have been productive elsewhere in society were converted to war use. Agriculture and other economic activities were greatly disrupted during the civil war. Politically, the war brought ethnic rivalry into the open. It also marked the beginning of the rise of the military with respect to the power and ability of that institution to chart a course for Nigeria. And with the oil boom following the civil war, the military was placed in the enviable position of controlling the enormous wealth of the country until 1979, when a civilian government was elected.

The Gowon regime, plagued by civil war, happened to be at the right place at the right time during the oil boom. Yet, its policies contributed to the scarcity of essential commodities in the midst of unlimited capital. Nigeria has never recovered from the goods scarcities of 1974. In the face of such scarcity, a "get rich quick" mentality developed. The next administration, that of Murtala Mohammed and Olusegun Obasanjo, came to power because the nation lacked effective leadership; orderly development was secondary. When the civilian politicians assumed leadership in 1979, they were not elected for economic development purposes, but to reestablish "democracy" in Nigeria. As *West Africa* (1984h:1394) points out:

> The very first deliberations in the various legislatures were on issues of personal remuneration and terms of service rather than on the more pressing issues of stabilising the nation's industry and agriculture. The irony is that the civilian politicians may have been trying to avoid hurting feelings when they ignored the clear signs of economic distress which they inherited from an era of economic extravagance generated by previous military governments.

So it appears that most of their efforts went to evolving and consolidating national political and economic structures rather than to promoting structural changes necessary for real social and economic development. The Buhari regime was the only government to come to power in order to restore the economy, and not because of political turmoil. Hence, the failure of that regime to turn the ailing economy around was stated as one of the reasons for its overthrow on August 27, 1985.

A second factor is the expectations generated in the 1966–79 period, during which most Nigerians believed theirs was a rich country. These expectations generated internal demands and pressures for economic self-sufficiency, modernity, industrialization, equitable distribution of income and national resources. The tremendous growth of road development contracts and import-substitution manufacturing left the country with huge debts and more dependent on the importation of raw materials, parts and components, and even food.

A third factor that acted to block Nigeria's development efforts is corruption. Nigerians knew that there was corruption from the highest level to the lowest level. Foreigners were also well aware. Headlines such as the following are common: "In Nigeria, Payoffs Are a Way of Life" (*Wall Street Journal* 1982); "Rooting out Corruption: Offering No Apologies, a New Leader Presses His War Against Indiscipline" (*TIME* 1984c:52). In describing the response to corruption during the Shagari regime, *West Africa* (1983a:191) points out:

> Unlike under the Murtala Mohammed regime, no senior government official has been sacked for corrupt practices. Instead, officials are redeployed and business goes on as usual. Several commissions of inquiry had been appointed in the past to investigate one fraud or another. In the end, things go the Nigerian way, as if nothing had happened.

Regarding the inability of President Shagari to do anything about corruption or to implement recommendations of the several commissions of inquiry, a columnist in the *National Concord* wrote:

> you would hand to the President a beautifully-bound copy of your . . . observations and recommendations. You can be sure that the President will shake hands and smile.
>
> He will even commend you for having contributed to our national development. But do not be disappointed if he subsequently sends the report to the National Archives where it will rest till eternity while we will continue business as usual.
>
> No honest man will blame him either. It is not within his powers, constitutional or otherwise, to reverse a trend whose symptoms appeared vaguely in the First Republic, and which the military fellows

brought to its peak, and the Second Republic appears to have hastened its climax. (*West Africa* 1983a:192)

Development plans often exhibited the political motives and self-interests of political leaders and public administrators rather than the goal of development per se. For instance, location of industries tended to be based on political rather than economic reasoning. Also, many projects were doomed from the start by their design. The reason was that a certain percentage of the cost of the projects was paid by the contractors to officials as bribes, commissions, or facility payments (Business International 1979:68). Not only that, but the materials and equipment used were frequently of substandard quality. If a project was successful, then the contract for that project could not be reawarded so as to collect more commission. But if the project failed, the contract could be reawarded and more bribes and commissions could be collected. All this, of course, meant more money in the foreign bank accounts of the privileged class, and less money for genuine national development.

Some may argue that the above analysis "blames the victim." External forces, for example, may have contributed to the corruption in Nigeria—especially through the MNCs. But they did not transform Nigeria into a corrupt nation. On coming to power, the Buhari regime declared war against corruption and indiscipline throughout Nigeria by means of a much publicized program christened W.A.I. (War Against Indiscipline). Since then, it has recovered large sums of money from the homes of former politicians—as high as N-3.4 million ($4.5 million) from the home of one former official. The "Dikko affair" in London ("the man in the diplomatic crate") is also illustrative. It involved an attempt to kidnap the former federal minister of transport, Umaru Dikko, and bring him back to Nigeria in July 1984. *West Africa* (1984j:1477) quoted the British *Daily Express* as saying that Dikko

. . . is a very rich man. He is one of around 30 top Nigerians sheltering in the West, most of them in Britain. Between them, it is estimated these have spirited out of Nigeria some 5,000 million [pounds sterling].

Another stumbling block to development in Nigeria has been the question of ethnicity and fear of domination. As was discussed in Chapter 4, the country known as Nigeria was a creation of Britain. Because of the heterogeneity in language, culture, background, life-style, and education, there is a tendency for certain ethnic groups within the federation to feel that they are being trampled upon by the major ethnic groups. Even among the major ethnic groups such as the Hausa, Ibo, and Yoruba, there has always been distrust. The job of nation building is made more difficult in such an environment. In response to ethnic rivalry, Nigerian regimes have sought to geographically balance economic development, with the consequence that location of projects has tended to be based more on political motives than on national objectives.

The above internal factors had serious impacts on Nigerian development strategies. Even if Nigeria had chosen different sets of development strategies, such internal factors would have impeded development efforts. The existence of the above internal factors underscores the seriousness of the Nigerian situation: If the country cannot control the factors in its internal environment, how can it be expected to control those in its external environment? Nevertheless, these factors cannot be taken as the source of Nigeria's crisis, but more nearly as its symptoms. Had Nigeria successfully established an autonomous development path, ethnic tensions might very well have eased, as demands for political autonomy could have been addressed within a context of secure progress. In turn, the assertion of military power might have been less likely, and the forces of corruption less powerful. Unfortunately, secure economic development did not come about.

WORLD MARKETS AND DEVELOPMENT IN NIGERIA

Ideology, the civil war, Nigeria's chosen development planning strategy, and the rise of the military have all affected Nigerian socioeconomic policy. Collectively, they constrained the possibilities for national development. However, as Nigeria came to discover, her socioeconomic progress was not a matter of internal economic activity and policy alone. The pursuit of unbalanced growth stressing first agricultural development for export, and then petroleum export and the infrastructural sectors, led Nigeria into an economic environment influenced significantly by circumstances beyond her control. She became dependent upon international supply and demand conditions that determined her export revenues, limited her ability to attract investment in sectors other than her growth sectors, set the terms of foreign exchange, and as a result determined her trade losses or gains. These factors are affected by events outside the control of Nigerian policy that have increasingly determined its success. As a result, Nigeria has been forced to balance internal demands for development with the external conditions of trade, international politics, internationalization of capital, and the availability of substitutes for Nigeria's export commodities.

Nigeria hoped to buffer these influences through its active membership in OPEC. By supporting a unified production and pricing policy, she sought to turn the world market to her advantage. OPEC was supposed to provide economic security for her petroleum economy, and in the process strengthen the chances for success of her unbalanced growth strategy. It briefly worked as Nigeria intended.

One of the most important external developments after the December 1973 price increases to work against Nigeria was the reaction of the oil importing industrial nations, particularly the United States. In February 1974, the Washington Conference on Energy was held, attended by 13 major Western industrial nations: Belgium, Canada, Denmark, France, West Germany, Ireland, Italy, Japan, Luxembourg, the Netherlands, Norway, the United Kingdom, and the United States (Mortimer 1984). All, except France, agreed to create a new organization to coordinate their energy policies. That organization, named the International Energy Agency (IEA), was inaugurated in November 1974.

The establishment of IEA was only a partial fulfillment of U.S. hopes. The United States was seeking a showdown between oil consumers and producers, while the other industrial members preferred U.N. initiatives based on broader issues than oil (Mortimer 1984). No doubt, the desire of France to create a network of European–Third World relations independent of U.S. influence was an important factor in turning Common Market partners away from confrontation. The United States favored an anti-OPEC coalition operating outside the U.N. institutions because the latter lent themselves to a show of Third World unity. While the United States failed in its effort to isolate OPEC countries, the incident underscored the fragile power Nigeria and others in fact wielded during the 1970s.

By the end of the 1970s, Nigeria's hoped-for OPEC buffer was under significant threat. Non-OPEC supplies were increasing rapidly as Britain, Norway, and Mexico offered petroleum at lower than OPEC prices. In 1982, the offer of non-OPEC petroleum by these countries began to put downward pressure on world prices. Cartel members saw an uneasy situation grow much worse. By 1983, members were ignoring OPEC pricing and quota policies, and the organization found itself unable to control its members. In October 1984, in response to cuts by Britain, Norway, and Mexico, Nigeria reduced its oil price to U.S.$28 per barrel. Later that year, when OPEC reduced its production ceiling, and in effect the production quotas of its members, Nigeria had to seek and obtain exemption from further reduction in its quota of 1.3 million b/d. In December 1984, Nigeria and Ecuador expressed their reluctance to support OPEC's plan for monitoring members' oil production levels, exports, and prices. Despite OPEC's struggle to maintain harmony among its members and Nigeria's crisis condition, one can expect Nigeria to remain a member of OPEC. However, the economic security sought through OPEC is no longer feasible.

Therefore, apart from the internal pressures with which Nigeria had to deal, there were also powerful external forces. These external factors contributed fundamentally to the development crisis now being experienced. The tremendous oil wealth of the 1970s flowed from the developed nations through Nigeria, but with its economy essentially colonial in structure, much of that wealth returned shortly to the developed economies to pay for her mounting import bill and MNC involvement in its development projects. Nigeria had little control over such factors. Her development planners and leaders should have anticipated the tactics of the various actors in her external environment and taken steps at least to reduce their impacts. But such actions could not have corrected Nigeria's basic problems, and cannot resolve them now. Without structural change in her political economy, Nigeria seems destined to drift from crisis to crisis.

CRITICAL DEVELOPMENT THEORY AND THE ISSUE OF STRUCTURAL CHANGE

Neoclassicism emphasizes the levels of capital accumulation and trade rather than political, social, and economic structure as the determinant of development. It in effect says that bigger piles of resources mean better conditions on the outcome side. It fails to take into account the national political, social, and economic structure that directs those resources to outcomes, and the international structure that directs resources to societies. Nor does it consider how much of the resources is actually left for development purposes after deducting for the leakages (such as corruption) embedded in the structure. Trade and capital accumulation, taken as the keys to development, have proved to be the shadow rather than the substance in the Nigerian situation.

This study has indicated that there has been an absence of structural change in Nigerian society since independence. The neoclassical development paradigm it has followed is almost entirely suited to reactive economic and political strategy. Neoclassicism counsels Nigeria to accommodate capitalism, even if it fails to produce or reinforce political and economic autonomy. In this respect, the paradigm has failed to instruct Nigeria on the central issue of structural change.

Despite these criticisms, the neoclassical paradigm is widely accepted and utilized by development planners in many developing countries and by international aid-giving agencies. The reason for continued use of neoclassical planning models was explained by Dudley Seers (1977:3) when he stated:

> Cultural lags protect paradigms long after they have lost relevance. The neoclassical growth paradigm has been remarkably tenacious—in fact it still survives in places. It has suited so many interests. It has been highly accepted to governments that want to slur over internal ethnic or social

problems. It has offered (not only in the hands of Walt Rostow) a basis for aid policies to inhibit the spread of communism . . . . It has provided international and national agencies with an "objective" basis for project evaluation, and goals for what should be called the Second Growth Decade. It has generated almost endless academic research projects and stimulated theorists to construct elaborate models. . . . Above all, as a paradigm it is very simple.

For Nigeria, the paradigm appears to have oversimplified the development problem.

If neoclassical development theory failed Nigeria on the key question of the necessary political, social, and economic structure for autonomous development, it is necessary to consider whether the ideas of critical development theorists would have been more appropriate in the design of her development strategy. As discussed in Chapter 3, critical development theories are distinctive for their analysis of the role of social, economic, and political structures in the development process. However, one of the major frameworks, Keynesian planning, is not relevant to the Nigerian context because it is applicable primarily to developed economies (see Chapter 3). Therefore, we will concentrate on the three critical development frameworks most appropriate to Nigeria: social development theory, neocolonial dependence theory, and the internationalization of capital framework.

Social Development Theory

Nigeria's leaders assumed that trade and capital accumulation were the fuels for the engine of growth and that these fuels would transmit the benefits of development to the rest of the economy (Cairncross 1961; Haberler 1968; Robertson 1947). The country also pursued a policy of unbalanced growth by investing autonomously in certain sectors that it considered most necessary to initiate a process of growth in trade and capital accumulation.

But social development theorists have argued that leading sectors seldom are successful in spreading their growth to the rest of the economy—particularly in a developing country like Nigeria, which must rely on raw materials sectors for trade. As Myrdal (1956b) argued in his "cumulative mechanism of causation" doctrine, as a mechanism international trade, by its very operation, had led underdeveloped countries to stagnation and impoverishment, and the developed countries into automatic cumulative growth. He also stated that international trade tends to have more "backwash" (unfavorable) effects than "spread" (favorable) effects. Also, according to Hymer's "law of uneven development," the system of MNCs has a tendency to produce poverty as well as wealth, underdevelopment as well as development, and to perpetuate existing patterns of inequality and dependency (Hymer 1972). MNCs prefer the path of "capital deepening" instead of "capital widening" in the productive sector of the economy. They do this by raising capital per worker through the installation of capital-intensive equipment and slowing down the expansion of the industrial labor force, thereby creating a dualism between a small, high-wage, high-productivity sector and a large, low-wage, low-productivity sector.

In the Nigerian context, there is merit to arguments both for and against international trade and capital accumulation as the keys to growth. That both can trigger growth has been in evidence in Nigeria, especially during the 1960s. But it is also true that leading sectors, particularly raw materials exports sectors, have failed to transmit their growth to the rest of the economy. Trade and capital have the ability to bring tremendous growth to a country's GDP and its foreign exchange reserves. But as Tables 7.4, 7.5, 7.7, and 8.1 show, only a few sectors grew, while imports of goods and services grew tremendously, during the period of oil boom, when trade and capital accumulation were at their peak. During the same period, agriculture declined, actually experiencing negative growth rate. No doubt foreign exchange ceased to be a constraint on growth during the oil boom, but it did not transmit its effects to the rest of the economy, particularly the agricultural sector. Inflation increased. Nigeria became a one-product exporting country. Neoclassical development strategy raised revenues in Nigeria but did not provide sustained growth, self-sufficiency, and economic independence. It also did not foster favorable structural changes in the social, political, and economic life of the nation.

But while the insights of social development theory on why Nigerian trade policy and capital planning failed are important, its prescriptions for what policies Nigeria should pursue in both areas are less helpful. Many of these theorists' recommendations are directed at the developed countries and their aid policies, matters over which Nigeria has no control. And the recommendations given to Third World planners presume either the existence of favorable aid and trade arrangements or a significant U.N.-style institutional apparatus directed by Third World countries. While Nigeria (and others) continues to advocate such things, she cannot wait for this new international order.

Neocolonial Dependence

The ideas of this school were applied to the African context by Dr. Kwame Nkrumah, the late president of Ghana. For him, neocolonialism meant that politically independent African states are independent only theoretically; in reality, these countries are still controlled by foreign countries (Nkrumah 1966). He believed that despite the decolonization in Africa, the colonial powers intended to keep the African economic and political structures dependent (Nkrumah 1963). According to Nkrumah (1970), a partnership had developed between the native bourgeoisie and the international monopoly in the post-colonial era. African governments, according to Nkrumah, had joined the MNCs in the exploitation of African workers and the rural proletariat; they protected MNCs from the resistance of the working class by banning and breaking strikes, and thus became the policemen for MNCs.

President Julius Nyerere (1974), in expressing the same view, compared neocolonialism to "independence for sale." The Arusha Declaration of February 5, 1967, had influenced policy makers throughout Africa. It called for the attainment of self-reliant nationhood, in which all economic activities would be controlled by the national governments (Nyerere 1968). To bring about self-reliance, neocolonial dependence theorists have called for African states to abandon trade- and industrialization-oriented development, and instead to build development through the promotion of agriculture and regional economic linkages.

In many respects neocolonial dependence analysis would seem to be accurate in regard to what has occurred in Nigeria. The role of foreign capital in Nigeria has certainly been substantial. A pattern of involvement, first in export agriculture and then in the petroleum industry, has been described in Chapters 5–7. Foreign capital was widely perceived as at least a partner in the corruption that has pervaded the Nigerian economy. MNCs were seen as the basic unit of imperialism in Nigeria and as responsible for the generation and perpetuation of corruption and underdevelopment, outflow of capital (that is, repatriation of profits to the investing nation rather than reinvesting the profit), debilitating export of the national economic surplus, and exploitative profit (Ejiofor 1976; Oni and Onimode 1975). MNCs are also believed to have impeded the economic and social development of Nigeria by controlling both the technology and the capital in the petroleum and export agriculture industries.

What these arguments fail to consider, however, is that we live in an interdependent world. Even such economic giants as the United States and the Soviet Union, to a certain extent, are dependent on other countries. For instance, the Soviet Union each year (except for the few years of grain embargo) buys millions of tons of wheat from the United States. The United States imports petroleum from other countries, such as Nigeria, Saudi Arabia, and Mexico. In addition, all industrial nations depend, to some degree, on imports for raw and intermediate materials. Neither Japan nor the European Community produces more than 25 percent of any of the minerals that are vital for its economy; even the United States produces less than 50 percent (Loup 1980:52; Noelke 1979). Perhaps those striving for self-sufficiency and economic independence ought to remember the words of Mahatma Gandhi:

> Interdependence is and ought to be as much the ideal of man as self-sufficiency. Man is a social being. Without interrelation with society he cannot realize its oneness with the universe or suppress his egotism. (As quoted in Attenborough 1982:87)

In assessing the neocolonial dependence theory, we need to recognize its strengths and its weaknesses. In the Nigerian context, its principal strength lies in the fact that the model recognizes the dependent structure of the economy despite political independence. Additionally, its focus on agriculture and regional economic linkages as bases of African development is important. Chapter 9 will provide further discussion on Nigerian agriculture. However, it should be noted that Nigeria, unlike many African states, has had a sizable industrial sector since midway through this century. It is essential that its development policy include an integrated urban/industrial and rural/agricultural orientation. Nigeria has been in the forefront of regional economic linkages. Its commitments in this area are illustrated by its relations with other West African countries that are members of the Economic Community of West African States. Its constitutional declaration to support Africa-wide development is indicative of the value placed on such relations. Finally, Nigeria's OPEC involvement also represents its commitment to Third World development needs. Despite these many efforts, however, Nigeria has not so far found in them the means to autonomous development. But by far, dependency theory's major weakness lies in the isolationist solution it proposes for Nigeria. Experience to date, including that of Tanzania, has shown that no nation can develop in isolation because of the interdependent nature of the modern world economy.

Internationalization of Capital

Regardless of whether Third World countries such as Nigeria can disconnect their economies from capitalist influence, reality compels Third World leaders and planners to recognize that their economies have been internationalized. Samir Amin and others have analyzed at great length the worldwide socioeconomic consequences of the integration of the peripheral economies into the international capitalist system. Their conclusion is that the underdeveloped countries are in a "blocked development" situation, due to their orientation to the world market and the consequent exogenous nature of their economies.

However, as discussed in Chapter 3, some proponents of the international capital perspective have recognized that dynamic elements exist in the changing historical conditions for capital accumulation in the Western countries, particularly since the petroleum price revolution of 1973. Increased differentiation among the countries of the Third World, according to this group, has been caused by the influx of European capital into certain countries with the intention of finding new areas of investment and new markets. Also, a few of these developing countries are believed to be in the process of establishing a basis for national capital accumulation. Furthermore, it is believed that the character of private direct foreign investment has changed. Foreign control over management has given way to participation in management, technical agreements, loans, production sharing, and supply contracts. The importation of large-scale producer goods complexes, often in the form of turnkey projects, has increased.

According to the internationalization of capital perspective, the international division of labor has changed, since 1973–74, from one in which the peripheral countries acted as raw material suppliers and markets for Western manufactured products to a more complex pattern. In recent years, there has been an increase in the export of industrial products from the developing countries to the industrialized countries. This supports the view that an export-oriented industrialization is taking place in some of the developing countries. Many of them succeeded in attracting foreign capital by providing a favorable investment environment: export processing zones with a well-equipped infrastructure, a highly qualified but low-paid (and unorganized) work force, tax exemptions, and other liberal investment regulations. But, as Samir Amin (1977:10) points out:

> Experience shows that participation of private or public local capital—however subservient—in the process of import-substitution industrialization is quite common. It also shows . . . that . . . it [is] possible to create a sector producing capital goods. The latter is frequently brought into being by the state. But the development of basic industry and a public sector does not in any way mean that the system is evolving towards a mature self-reliant form, since the capital-goods sector is here used, not for the development of mass consumption but to serve the growth of export and luxury-goods production. So this . . . phase of imperialism is by no means a stage towards the constitution of a self-reliant economy.

The internationalization of capital school deserves credit for including the shift in international capital structure in its analysis. But the Nigerian development patterns have not supported the idea that the development of export industries and a public sector would necessarily lead to an autonomous and self-sufficient economy. Rather, the Nigerian situation supports Samir Amin's arguments quoted above. Moreover, Nigeria's structural crisis, despite its ability to use petroleum as a means for becoming a circuit in the international circulation of capital, undercuts some of the analyses of this school. Realizing the international character of capital and its control by the major capitalist societies and organizations has yielded valuable insights on the development process. Nigeria has been one of the few Third World countries able to generate substantial domestic capital, yet it has been unable to produce autonomous development.

SUMMARY

In this chapter, a critical analysis of the Nigerian development experience and the economic and sociopolitical crisis that has accompanied the oil boom was provided. It was established that the ideology of nationalism, the choice of the neoclassical paradigm, and adoption of the single-sector planning approach contributed to the failure of Nigeria to create a self-sufficient, autonomous, and independent development process. Also examined were the roles of certain internal and external forces in Nigeria's development experience; it was indicated why and how they contributed to the problem. However, the analysis has led to the conclusion that the greatest obstacle to autonomous development has been Nigeria's structural condition of underdevelopment. This condition derives from its colonial past, which created an industrial and export agriculture economy dependent upon foreign capital and trade. The decision by Nigeria's leaders and planners to orient post-colonial development toward a petroleum economy only reinforced the structural condition of neocolonialism.

While critical development theorists can be seen to have anticipated the problems Nigeria experienced, their proposals to address the need for change in social, political, and economic structures were found to offer only modest hope. Both neoclassical and critical development theories fail a country like Nigeria for a similar reason. As noted earlier, the neoclassical paradigm urges a developing country to accommodate capitalism, even though such a strategy is often in conflict with the desire for autonomous development. In this respect, the paradigm treats the problem of development as a reaction to capitalism. While critical development theory is more accurate in its portrayal of the implications of capitalist development for developing countries, it also instructs countries about planning strategies from the vantage point of external capitalist conditions. Neither tells Nigeria what it needs to know: how to effect change in its political, social, and economic structures so as to make autonomous development possible.

What, then, are the lessons for Nigeria and other developing countries? This is the subject of Chapter 9.

9 Breaking the Condition of Underdevelopment

> Time and circumstances invariably lead all nations into revolutionary periods wherein they must make profound alterations or perish. New conditions and new attitudes put so much pressure upon established values and institutions that fundamental changes that must be more than fine tunings of the status quo become necessary. Sometimes such revolutions rip societies to shreds and completely new political forms rise from the ruins. Other times, the result of revolution is an amalgam of elements of the old institutions fused with elements of the new. This synthesis alters institutions so that they are both sufficiently familiar to keep the society functioning, yet modern enough to respond to contemporary demands. (Scott and Hart 1979:207)

INTRODUCTION

As already argued in this book, the lack of development in Nigeria, despite its petroleum-revenue-stimulated growth of the 1970s, is a result of a structural condition of underdevelopment. It was also argued that Nigeria remains a neocolonial society, that is, one in which its social, political, and economic character resembles its colonial past. It remains one that depends on, and is indirectly controlled and influenced by, developed economies through its dependence on international trade and foreign capital attraction. In order to break this structural condition of underdevelopment, Nigeria must undergo structural change that will create conditions for autonomous development, that is, development that is internally controlled, directed, and influenced. The result of autonomous development will be the achievement of economic and political self-reliance for Nigeria.

What are some concrete steps that Nigeria needs to take in order to achieve autonomous development? In discussing what the plan of action should be, I have decided to stay within the internal environment of Nigeria. That is, the recommendations below are based on those issues that Nigeria has control over and can do something about. In Chapter 8, we examined the external forces that Nigeria has to contend with. It has little or no control over those external forces. Also, the international environment is so dynamic that today's prescriptions may no longer be effective tomorrow. For a country such as Nigeria, however, the issue posed by its external environment cannot be addressed until internal change is effected. This is the lesson of Nigeria's largely unsuccessful indigenization policy. Therefore, the recommendations that follow are discussed under three broad headings: economic, sociopolitical, and planning.

RECOMMENDATIONS FOR ACHIEVING ECONOMIC SELF-RELIANCE

Development implies major structural changes and correspondingly large modifications in social and institutional conditions of a country. Development is a condition of socioeconomic and political wellbeing and autonomy, the opposite of underdevelopment. It is possible, however, to have development within underdevelopment, where growth is taking place in a country showing the negative characteristics of underdevelopment. Large modifications in the social and institutional conditions of a nation are prerequisites to development (Gill 1963).

Stressing the importance of human and social factors in the development process, Schiavo-Campo and Singer (1970:6) argued:

[D]evelopment relates to something more fundamental, which may be defined as the establishment of a mechanism which will produce self-sustaining and cumulative indigenous economic improvement. The essential problem of economic development is not production, but the capacity to produce. However, the capacity to produce is clearly not determined exclusively by the stock of physical capital or of natural resources; a vital component of it is inherent in people. This leads, of necessity, to the relevance of the human and social factors in development.

As Nigeria has discovered, large sums of money alone cannot transform a country into a developed, self-reliant nation. It helps to have capital, but that alone cannot guarantee autonomous development. This does not mean that Nigeria should isolate itself from the world economic systems. Rather, it means that Nigeria should strive to achieve a new balance between its internally directed production system and social needs, and the international economic order. This must be done in such a way that the provision of basic goods and services is not threatened by external economic and political forces. The following are some of the priority areas for Nigeria.

Agriculture

One of the areas in which Nigeria should strive for self-sufficiency is food production. This entails paying more than lip service to agriculture. It requires concrete action rather than the type of publicity blitz that accompanied "Operation Feed the Nation" and similar agricultural schemes. It requires a change in the traditional attitude of planners and policy makers, who have ignored agriculture because they believe that it "deprives development of dramatic character and makes it appear rather pedestrian" (Alpert 1963:173).

In Nigeria, agriculture had appeared to be less glamorous, not only to planners and policy makers, but also to public and private lending institutions. The Nigerian Agricultural and Cooperative Bank (NACB) was established by the federal government in 1973 to finance agriculture and agro-based industries. The bank was to make loans to individual farmers either directly or indirectly through state governments and cooperative associations (Nweze 1980:255–62; Okafor 1980:140–42). Between 1973 and 1976, NACB approved 99 projects for a total of N-103 million. Out of these projects, 76 received disbursements that amounted to only N-22.8 million, a 22.1 percent disbursements/approval ratio (Okafor 1980:143).

An analysis of commercial bank loans and advances, prescribed and actual, by classification for 1972 to 1978 shows that while the rest of the Nigerian economy received substantial loans and advances during the oil boom, investment in agriculture was deliberately kept to a minimum. On the average during those years, the Central Bank of Nigeria, as part of the government's monetary policy, prescribed to commercial banks that 5.14 percent of all commercial loans and advances should be awarded to agriculture. By contrast, during the same period, the average prescription for manufacturing was 30 percent. The actual performance—that is, actual loans and advances to agriculture—was even lower—it averaged only 3.26 percent of all loans and advances, while manufacturing received an average of 27.18 percent (Nwankwo 1980). This trend needs to be reversed in order to bring Nigeria back to its historic self-reliance in food production.

While developed countries subsidize their farmers and even pay them for not producing, Nigerian farmers are penalized for producing. Even though the state marketing boards have been replaced by a federal board for each commodity, it still does not pay to farm when one can become a middleman. Farmers ought to be given a prominent role in the development process. Their needs and their problems must be addressed in national development plans, loan priorities, and infrastructure decisions. For this to occur, an accurate census is needed to determine how many farming families there are in Nigeria, their farming methods, the crops they grow, and the problems they face. The importance of accurate demographic information cannot be overemphasized. It is essential for planning, not only in the public sector but also for industries. Nigeria has been using the 1963 census, compounded at 2.5 percent per annum. The accuracy of the 1963 census is questionable and is only made worse by the assumption that growth has been constant at 2.5 percent. Without a population census, planning in Nigeria cannot make efficient use of the human resources available to the country.

Integration of Agriculture and Industry

In order to effect structural change, the integration of the agricultural and industrial sectors will be necessary. This will entail paying attention to the linkages between the two sectors, and how they support each other. Raising productivity in both sectors is essential. In their development efforts, the neglect of agriculture has been a common mistake made by most Third World countries, including Nigeria. The shortage of food, manufacturing materials, foreign exchange, and other products and services resulted from strategies that supported the dependent structure of neocolonial Nigeria. In this structure, industry and agriculture have very few backward and forward linkages. There was little attempt made to develop a technology that utilizes the natural factor endowments of the country. Thus labor, the factor of production that is available in the greatest quantity in Nigeria, has been inefficiently utilized. Rather than continue to concentrate on capital-intensive sectors, which have left the country with frustration and greater dependence, it is time to reevaluate the entire industrial sector. This involves determining what Nigerian society actually needs as against what it wants (demonstration effects). Nigeria must then develop local factors of production necessary for producing the goods and services that will satisfy such needs.

At the same time, all import-substitution industries should be evaluated with a view to establishing their contributions to the goal of a self-reliant Nigeria. In fact, where all the materials for an industry are imported (and in many cases overbilled), it may be better for Nigeria to import the finished product rather than assemble it locally. As an example, one may ask whether the assembly of motor vehicles in Nigeria has resulted in the transfer of technology and thus made Nigeria less dependent on the developed economies, or whether it has increased Nigeria's dependent condition. We may also ask whether it costs the Nigerian consumers more to buy "assembled in Nigeria" products than to import the same products (which in many cases are of higher quality). While the intentions of the government to control its foreign exchange reserves are noble, the strategy of import-substitution industrialization is notorious for ensuring the failure of such intentions.

Electricity and Communication

In the 1950s, Chief Obafemi Awolowo wrote in his memoirs that without electric power, a potable and reliable water supply, and efficient means of communication, economic development would not be well founded (Awolowo 1960:285). These necessities are still lacking in Nigeria, and development continues to be lacking as well. At independence, it was necessary for the federal government to control the ownership and distribution of electricity. The same applied to telephones and telegraphs. Over a quarter of a century later, it is time to hand over these two areas to state and local governments and private enterprises.

To proceed, the federal government should consider allowing each local government area to choose the supplier of electricity and communication facilities. This means that private companies, in addition to state government-owned and local government-owned corporations, would be allowed to bid for each local government area. Nigerians proved during the indigenization exercise that they are willing to invest and have the funds to support that willingness. Each state government can then establish its own utility board to regulate the utility companies and coordinate their activities.

By handing over the determination of who generates and distributes electricity and who provides telephone/telegraph services at the state and local levels, the central government will be doing itself a favor—it will have more money in its treasury because it will not be paying out subsidies. It will be able to generate revenues by collecting taxes on the sale of electricity, and telephone and telegraph services throughout the country. Rather than continue to concentrate infrastructure on a few primary cities, a broader-based development process needs to be established. Decentralizing development throughout the country will enable infrastructure to be distributed in such a way as to prevent people from gravitating toward the main cities. In this way, the needs of agriculture will be better served.

Development from the Local Level Up

As part of breaking the structure of underdevelopment, development should be allowed to spring up from the local level rather than trickle down from the central government. The seeds of development, such as capital accumulation, skilled manpower formation, and technological progress, ought to be sown at the local and state levels, rather than at the federal level. The notion that the federal government will transplant the fruits of development, such as lower unemployment rate, higher literacy rate, better nutrition, lower death rate, industrialization, and a large consumption basket to the state and local governments has not materialized despite efforts in this direction since 1960. While it is thought to be easier to transplant the fruits than to plant the seeds and wait for their growth, it is doubtful that such a practice will yield long-term development. Automatic transfer of fruits of development has a way of slowing down the development process. In Nigeria, the central government has created demands for commodities manufactured by factories using imported materials and parts, but in so doing it has reduced the desirability of farm work and the value of its products. This has been a significant source of the country's foreign exchange problems.

Summary of Economic Recommendations

The above recommendations do not represent all that could be done for the Nigerian economy. Others include the systematic reduction of the country's reliance on petroleum revenues as a main source of government revenues and using the oil money to develop other resource bases in order diversify the country's economic base. Both of these matters are discussed more fully later in this chapter. Also, import tariffs need to be restructured in order to encourage investment in agriculture and increase local production of food (among other reasons).

In order to effect autonomous development in Nigeria, it is necessary to expand and strengthen its internal economic base. This involves creating a condition whereby there are forward and backward linkages between the various sectors of the economy. This also entails ending the duality between agriculture and industry. It is essential that Nigerians should trade with Nigerians and produce for Nigerians. That is the reason for the focus on agriculture and the linkages between agriculture and industry.

SOCIOPOLITICAL RECOMMENDATIONS

Nigeria seems to be in a cycle of political failure: it started at independence in 1960, with a civilian government; in 1966, two military coups; in 1975, another military coup; in 1976, a military coup, followed in 1979 by a return to civil rule. On New Year's Eve 1984, the civilian government was overthrown in another military coup; and on August 27, 1985, another military coup took place. The fact remains that the Nigerian political, socioeconomic, and administrative systems were not conducive to genuine national development. There were just too many obstacles embedded in them. Political development has not taken place in Nigeria because its political systems have not been released from the shackles of the structure of underdevelopment. Without political development accompanying economic development, national development in Nigeria will continue to be elusive.

A major issue that Nigeria needs to address is that of governance. How can it become truly governable as one nation? Nigerians did not decide to establish a country named Nigeria, with all its ethnic groups and diversified cultural groups: the British did. However, since Nigerians find themselves in one country and have fought the War of National Unity (July 1967 to January 1970), Nigeria ought to remain one country and build a nation based on a choice among the alternatives depicted in Figure 9.1.

FIGURE 9.1   Options for Intergovernmental Relations in Nigeria

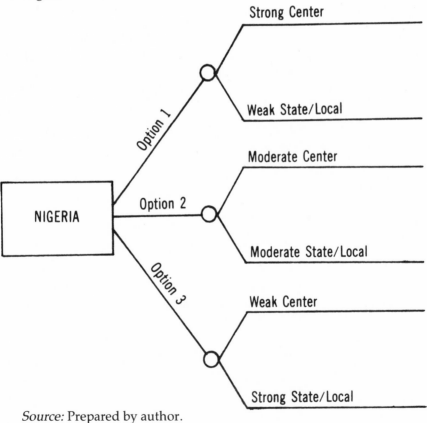

Source: Prepared by author.

Option 1 operates on the basis of a strong center and weak state/local governments. To date, this option has met with little success in Nigeria. Option 2 may be a better choice because it seems to balance out the power, but in a federal arrangement such as that of Nigeria, the concentration of power tends to tilt toward the center. Therefore, Option 3, which requires a weak center and strong state/local governments, seem to be the best choice for a "new" Nigeria, and should provide the political stability that Nigeria has lacked since its independence. One of the underlying reasons for political instability is that the center has attracted so much power and attention. This has strangled healthy competition among states for business and for development. If everyone in Nigeria becomes aware that power is concentrated at the state and local levels, then the state and local governments can concentrate on their own affairs, realizing that if there is going to be any development at all, such development must spring from the local level. The seeds of development must be sown at the local level rather than being transplanted from the central government. Nigeria is a vast land; it is only the local people who can develop themselves.

Under Option 3, the role of the central government should be redefined with a view to reducing it and transferring the appropriate tasks to the state and local governments. Nonetheless, it is recognized that certain functions, such as defense and security, require centralization. The idea is to have the central government execute those functions that must be centralized in a multistate "new" Nigeria. Likewise, the allocation of resources should be commensurate with responsibilities. Option 3 is recommended in order to effect political development in a "new" Nigeria.

This option will alleviate the fear of certain sections of the country that the central government has continually been dominated by certain parts of the country. As shown in Table 9.1, during Nigeria's political life from 1959 to 1985, the central government was led by a Nigerian of northern origin for 22 years; leaders from the other areas have had the opportunity to rule the country for 4 years—only as military officers. This domination has become an issue only because Nigeria has maintained a strong central government. Therefore, this issue needs to be dealt with honestly if there is to be peace and harmony in a "new" Nigeria. When no areas of the country feel oppressed by other areas or ethnic groups, it will not be difficult for each citizen to see himself or herself as a Nigerian first, and to recognize the country as just, equitable, and able to protect the collective interests of all people. At that point, there will not be a necessity for political quotas. For Nigeria, this will come about only if the central government is relatively weak and more autonomous state governments are established. As Hatch points out, it is time for patriotic Nigerians to tackle this problem by surmounting ethnic allegiances:

TABLE 9.1. Distribution of Central Government Leadership, 1959–85

Leader	Type	Period	No. Yrs.	North	South
Balewa	civil	1959–66	7.0	7.0	—
Ironsi	military	1966	0.5	—	0.5
Gowon	military	1966–75	9.0	9.0	—
Mohammed	military	1975–76	0.5	0.5	—
Obasanjo	military	1976–79	3.5	—	3.5
Shagari	civil	1979–83	4.0	4.0	—
Buhari	military	1984–85	1.7	1.7	—
Babangida	military	1985–	0.5	0.5	—
Totals	civil		11.0	11.0	—
	military		15.7	11.7	4.0
	all		26.7	22.7	4.0

Source: Provided by author.

The effects of the 1914 decision have been tragically evident ever since. Yet it would be unctuously patronizing to suggest that the Nigerian crisis has been thrust upon helpless Africans by Britain. African missionaries, traders, and other professionals urged the British government to take control of their country in the late nineteenth century. None of them had the foresight to see that only the establishment of separate states, or a deliberate policy of amalgamating administrations and increasing contact between ethnic groups, could avoid a conflict of nations within a single state system. Nor have Nigerians caught up in this crisis been able to summon up the intellectual honesty needed to recognize the reality of the situation they have inherited and, surmounting ethnic allegiances, meet the challenge of choosing either separate nations or a genuine multinational state. (Hatch 1970:239–40)

This question, though sensitive, must be resolved in order to bring political stability and national cohesion to a country, such as Nigeria, where ethnic, linguistic, demographic, religious, and other differences exert simultaneous pressures on its political life.

In addition, for Nigeria to return to a truly democratic system of government, its election laws should be reevaluated. It has always been required that candidates running for office resign their public appointments. That law discriminated against and prevented the not-so-rich and those who could not afford to be self-employed, but who are potential public-service-oriented persons, from running for office. What the requirement meant was that one could lose a race for office and also lose one's means of livelihood. It was an indication that the nation was unwilling to invest in potential leaders or that the law was simply an instrument for allowing only rich, self-employed citizens to run the government. The strategy was part of the neocolonial attitudes that Nigeria needs to change. Aristotle wrote:

> If liberty and equality, as is thought by some, are chiefly to be found in democracy, they will be best attained when all persons alike share in the government to the utmost. (As quoted in Reedy 1984:29)

Furthermore, one of the conditions laid down for the establishment of political parties before the 1979 elections was that the party must have national features. On the surface, it sounded objective and nationalistic, but realistically it created problems. A major problem was that the only people who could organize "national" parties were the wealthy and established politicians. So it turned out that the electoral laws, which were supposed to further the democratic process, became a tool for eliminating newcomers and potential leaders from the political arena. What a "new" Nigeria needs is a system that guarantees equal access to all citizens.

These recommendations do not in any way constitute a complete list of the solutions to all the problems in Nigeria's political system. They only represent a good starting point in a long struggle. In short, political structure in a "new" Nigeria should foster autonomous development. To do this will require a greater degree of participation at the state and local levels. Only in this way can development occur from the local level up.

RECOMMENDATIONS FOR NATIONAL PLANNING

The problem with national planning in Nigeria is that the planning apparatus has focused on development principles of foreign trade and capital accumulation, and has channeled development through a single sector. To encourage autonomous development, Nigerian national planning must adopt different principles and different approaches. The guiding idea of a national plan should be to promote linkages between rural and urban, and agricultural and industrial, communities, and to increase the diversity of production and trade among Nigerians. To do so will require abandonment of the single-sector planning approach and, in its place, the use of multisector planning in which projects are related to the growth of several sectors and to interrelationships among sectors. Even with a more accurate census, Nigeria will not be able to plan from the vantage point of input–output analysis or other sophisticated techniques. But it can, and must, improve the participation of all its economic sectors in establishing planning goals and activities. To do this will necessitate a decentralization of the planning function while, at the same time, ensuring the consistency of national goals.

Therefore, it is recommended that a Bureau of National Development be established. This office is not a planning office in the conventional sense, but a coordinating office for national development plans. The major ministries, such as Agriculture, Finance, Energy, Transport, Trade and Industry, Education, Health, Telecommunication, the Central Bank, and the Petroleum Corporation, will have representation on the board of the bureau. The chairman will be the head of state or, in his absence, his deputy. The main objectives of the board are to ascertain that development goals are achieved, to control development efforts, and to keep the country moving in the desired direction. The idea is to coordinate the development activities and individual projects so as to increase internal consistency and minimize conflicts. The secretariat of the bureau will be headed by a secretary general, who will oversee its day-to-day running. His staff will consist of development analysts, liaison officers, and general administrative staff. Figure 9.2 depicts the structure of the bureau.

The implementation of this recommendation will necessitate the creation of state development offices and local area development offices, as shown in Figure 9.3. In their planning efforts, the state development offices will coordinate the efforts of the local development offices in their respective states, while the Bureau of National Development will be responsible for coordinating the efforts of state development offices. In this way, planning for development will spring from the local level rather than being transplanted from the federal level. By following this arrangement, each local government area does its own planning; then the board as a body, after due consideration and evaluation of backward and forward linkages, and on the basis of available resources, will recommend combinations of main-sector development projects.

The Bureau of National Development will be unique in its basic function: It will deal only with national development rather than with the day-to-day running of the country or with recurrent expenditure, both of which usually prevent enough time and resources from being devoted to national development. Also, since a strategy indicates the route chosen to reach the desired long-run objective, it is important and necessary that the preferred development strategy be made explicit to the representatives of the various sectors concerned. This is essential in order for its social and economic implications to become evident, particularly if it involves structural changes that are induced by governmental policy. At the local and state levels, representatives of agriculture, industries, labor, and education should be invited to participate in the planning process.

A problem that will need to be addressed at all levels of the planning apparatus is that of bureaucratic red tape—a carryover from the colonial administration. These numerous bureaucratic obstacles were designed to strangle the Nigerian development efforts. After independence, these practices have continued to be widespread in most government offices. The author recalls what many businessmen had to go through in order to obtain simple administrative services, such as registration of a business name or change of a business address. Perhaps that explains why most businessmen in Nigeria prefer to be middlemen rather than producers or manufacturers; the trouble may just not be worth it. In a "new" Nigeria, public officials need to be service-oriented, and must realize that they are contributors to the goal of autonomous development.

Above all, Nigeria's broad planning goal should be domestic prosperity, political stability, and the expansion of her internal economic base. This is crucial to turning to her advantage the opportunities offered in an interdependent world.

LESSON FOR NIGERIA AND OPEC: THE LIFE CYCLE OF PETROLEUM AS AN EXPORT COMMODITY

The concept of life cycle is not novel; it has been used as "product life cycle" in the field of marketing (Levitt 1966) and as "project life cycle" in the capital budgeting aspect of financial management (Brigham 1979).

Petroleum, as an export commodity, has a life cycle that can be divided into four stages: development, growth, maturity, and saturation/ decline. Figure 9.4 relates these stages to the case of Nigeria. The developmental stage covered the period between 1958 and 1963, when petroleum was still a small industry completely controlled by Shell-BP, with all the crude oil exported to Britain and Holland. The growth stage, which covered 1964–69, witnessed an expansion of the market for Nigerian crude oil. During that period, Nigeria's customers included West Germany, France, Canada, the United States, Argentina, and Ghana. The maturity stage (1970–80) saw the addition of the Japanese market, and it was during that period that the United States emerged as Nigeria's largest market. During this period, the United States imported roughly 30 percent of Nigerian crude oil. Nigeria also supplied the West African market, which included Benin, Chad, Ghana, Niger, and Togo. The decline stage started in 1981, when production dropped to as low as 0.6 million b/d. In 1984, Nigeria maintained an OPEC production ceiling of 1.3 million b/d. On October 29, 1984, OPEC oil ministers decided to reduce the OPEC produc-

FIGURE 9.2 Level 1, Proposed Structure for Bureau of National Development Planning

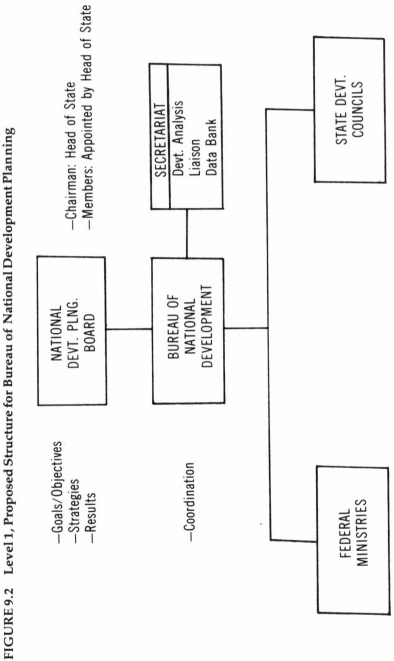

—Goals/Objectives
—Strategies
—Results

—Chairman: Head of State
—Members: Appointed by Head of State

NATIONAL DEVT. PLNG. BOARD

BUREAU OF NATIONAL DEVELOPMENT

SECRETARIAT
Devt. Analysis
Liaison
Data Bank

—Coordination

FEDERAL MINISTRIES

STATE DEVT. COUNCILS

Source: Prepared by author.

FIGURE 9.3 Level 2, Proposed State/Local Development Planning Apparatus

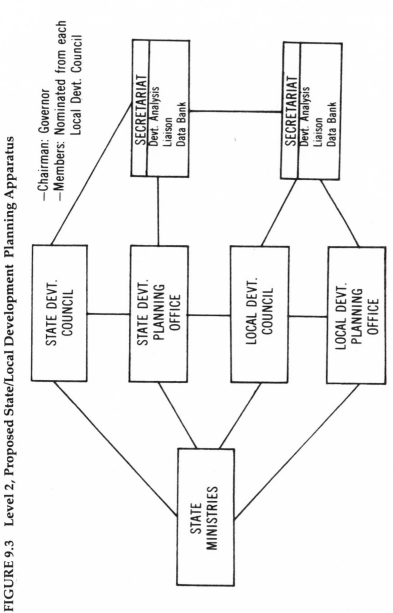

Source: Prepared by author.

173

FIGURE 9.4 Life Cycle of Oil in Nigeria (1958–84)

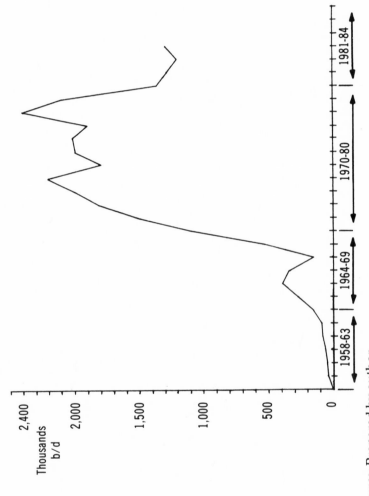

Source: Prepared by author.

174

tion ceiling of 17.5 million b/d by 1.5 million b/d, effective November 1, 1984. All members were expected to share in the production cut in order to maintain the OPEC price level. Nigeria was later exempted from participating in this cut.

The importance of understanding this concept of the life cycle of oil as an export commodity is that each stage requires different strategies. The crucial stage for Nigeria was the maturity stage. During that period, the country had capital to diversify but failed to do so. When prices and revenues were high, there was less incentive to diversify; when the oil revenues were drying up, there arose greater incentive to diversify. When there was a higher incentive to diversify, the problem became lack of capital: All the money had been spent on building roads and a new capital city (apart from the portions mishandled). The problems of a Third World country such as Nigeria are further compounded at such a time, because in a time of difficulty it can hardly count on either the IMF or the World Bank to lift it out of its economic problems. During Nigeria's negotiation for an IMF loan in 1984, some of the IMF's requirements were devaluation of the currency (naira); cutting petroleum subsidies; and trade liberalization. The consensus in Nigeria was that these conditions were detrimental to the already ailing economy. Labor leaders warned that the IMF conditions would lead to higher prices for many consumer goods and services, such as gasoline, electricity, and transport, as well as to the closing of more factories and greater foreign domination of the economy (*South* 1984a:24). It was believed that the poorest people were likely to be hit hardest by the IMF requirements.

From the above, it becomes clear that when the need is there to diversify, and there is little or no capital left to execute the diversification plan, international aid will be lacking. Even where such aid is available, the strings attached, if implemented, could throw a country into chaos.

Had Nigerian planners recognized the stage in which Nigerian petroleum export was in 1980–81, they would not have based the resources to implement the Fourth National Development Plan (1981–85) on the production and sale of 3.0 million b/d of oil. If planners had realized that oil-importing nations were creating temporarily high demand while stockpiling for strategic reasons, they would have forecast lower future demand leading eventually to excess supply and depressed prices. Nigeria ought to have remembered that this same strategy was adopted by cocoa-importing nations to depress the price of cocoa.

The concept of life cycle should be applied to any export commodity, particularly raw materials, that a Third World country relies upon for its foreign exchange. By so doing, a country will be able to adjust its strategies so as to maintain conditions contributing to its self-reliance and self-determination.

In the case of Nigeria, one thing is certain: Nigeria is now in the saturation/decline stage in the life cycle of its petroleum as an export commodity. It is time to diversify its economic base so as to replace petroleum as the main source of government revenues. Also, revenues from oil should be utilized to develop other productive sectors rather than being invested in infrastructure. In addition, the petroleum sector should be thought of as a sector that can support domestic production. An example of how it can do this is provided by natural gas, a by-product of crude petroleum, which currently is being flared in Nigeria. If appropriate techniques are developed to recover natural gas, it will form part of the diversification program and also make natural gas available for internal use, both private and industrial.

CONCLUSION

The above recommendations will not resolve the problem of the intrusion of world markets into Nigeria's sociopolitical and economic life. Nor do they provide the basis for Nigeria to uncouple from aspects of the world market system that are deleterious to its interests. Instead, these recommendations focus on effecting internal change in the belief that Nigerians must first understand and act on the problem of national economic and political autonomy. Internal change must occur before they can begin to address the issues of external influence.

Decentralization is advocated at all levels of Nigerian society, in an attempt to bring Nigerians into dialogue about the problems they face. It is an attempt to reeducate them about the need for, and obstacles to, autonomous development.

Economic and sociopolitical development failed to take place in Nigeria because of a structural condition of underdevelopment inherited at independence and continued by Nigeria's development strategies. These strategies relied on trade and capital accumulation, making Nigeria dependent upon external economic conditions. The cycle of underdevelopment must be broken, which will require considerable internal change.

The road ahead for Nigeria is likely to be a difficult one. Life will be harder before it is easier. Long-term goals and objectives, rather than short-term ones, will have to be emphasized. Nonetheless, Nigeria has all the human and nonhuman resources it needs to become a great nation. However, unless it undergoes structural change, whatever is done to "patch" the society will only ameliorate the present situation without removing the real problem facing Nigeria. Sincere and honest reevaluation of the country's entire political, socioeconomic, and planning systems is required. It can be done. Development does not have to be elusive.

Selected Bibliography

Aboyade, Ojetunji. 1966. *Foundations of an African Economy: A Study of Investment and Growth in Nigeria.* New York: Praeger.

_____. 1968. "Relations Between Central and Local Institutions in the Development Process." In Arnold Rivkin (ed.), *Nations by Design.* Garden City, N.Y.: Doubleday.

Adedeji, Adebayo. 1969. *Nigeria Federal Finance.* New York: Hutchinson Educational.

Adedeji, Adebayo (ed.). (1973). "Management Problems of Rapid Urbanization in Nigeria. The Challenge to Government and Local Authorities." *Report of the National Conference on Local Government, Sokoto, 5–8th January 1972.* Ile-Ife: University of Ife Press.

Africa Guide 1983. Essex, England: World of Information, 1983.

Africa Magazine. 1981. 116 (April):76–91. "Nigeria 4th National Development Plan 1981–85."

Africa South of the Sahara 1983–84. 1983. London: Europa Publications.

Agarwala, A.N., and S.P. Singh. 1963. *The Economics of Underdevelopment.* New York: Oxford University Press.

Ake, Claude. 1981. "Presidential Address to the 1981 Conference of the Nigerian Political Science Association." *West Africa* (25 May):1162–63.

Akeredolu-Ale, Ekundayo O. 1975. *The Underdevelopment of Indigenous Entrepreneurship in Nigeria.* Ibadan: Ibadan University Press.

_____. 1976. "Private Foreign Investment and the Underdevelopment of Indigenous Entrepreneurship in Nigeria." In G. Williams (ed.), *Nigeria. Economy and Society*: 106–22. London: Rex Collings.

Akinsanya, Adeoye. 1983a. *Economic Independence and Indigenisation of Private Foreign Investments: The Experiences of Nigeria and Ghana.* New York: Praeger.

_____. 1983b. "State Strategies Toward Nigerian and Foreign Business." In I.W. Zartman (ed.), *The Political Economy of Nigeria*:145–84. New York: Praeger.

Alam, M. Shahid. 1982. "The Basic Macro-economics of Oil Economies." *The Journal of Development Studies* (January):205–16.

Alpert, P. 1963. *Economic Development.* New York: Free Press of Glencoe.

Aluko, I.S. 1972. "Nigeria. Toying with Nationalism." *Africa* no. 10.

Aluko, Olajide. 1981. *Essays on Nigerian Foreign Policy.* London: George Allen & Unwin.

Aluko, S.A. 1973. "Industry in the Rural Setting." In *Rural Development in Nigeria*. Ibadan: University of Ibadan Press, 1973.

Amin, Samir. 1970. *The Maghreb in the Modern World: Algeria, Tunisia, Morocco*. Baltimore: Penguin Books.

_____. 1973a. *Neo-colonialism in West Africa*. Harmondsworth, England: Penguin Books.

_____. 1973b. "Underdevelopment and Dependence in Black Africa." *Social and Economic Studies* 22 (March):177–96.

_____. 1974a. "Accumulation and Development: A Theoretical Model." *Review of African Political Economy* 1 (August–November).

_____. 1974b. *Accumulation on a World Scale: A Critique of the Theory of Underdevelopment*. New York: Monthly Review Press.

_____. 1976a. *Imperialism and Unequal Development*. New York: Monthly Review Press.

_____. 1976b. *Unequal Development: An Essay on the Social Formations of Peripheral Capitalism*. New York: Monthly Review Press.

_____. 1977. "Self-Reliance and the New International Economic Order." *Monthly Review* (July/August).

Anchishkin, A. (ed.). 1980. *National Economic Planning*. Moscow: Progress Publishers.

Arnold, Guy. 1977. *Modern Nigeria*. London: Longman Group.

Arusha Declaration. 1967. Dar-es-Salaam, Tanzania: Government Printer.

Ashley, W.J. (ed.). 1929. *Principles of Political Economy*. New York: Longmans, Green.

Associated Press. 1984a. "Oil Boom's End Leaves Nigeria over a Barrel." *Sunday News Journal* (January 8): C5.

_____. 1984b. "Nigeria Rulers Decide to Get Tough." *The Morning News* (May 14).

_____. 1984c. "Nigeria Trims Its Oil Prices." *The Morning News* (October 19):A1.

Attenborough, Richard. 1982. *The Words of Gandhi*. New York: Newmarket Press.

Awolowo, Obafemi. 1960. *AWO. The Autobiography of Chief Obafemi Awolowo*. Cambridge: Cambridge University Press.

_____. 1968. *The People's Republic*. Ibadan: Oxford University Press.

Ayida, A.A. 1971. "Development Objectives." In A.A. Ayida and H.M.A. Onititi, (eds.), *Reconstruction and Development in Nigeria*. Ibadan: Oxford University Press.

Bagu, Sergio. 1949. *Economia de la sociedad colonial.* Buenos Aires: Libreria "El Ateneo."

Balabkins, Nicholas. 1982. *Indigenization and Economic Development. The Nigerian Experience.* Greenwich, Conn.: JAI Press.

Balewa, Abubakar Tafawa. 1964. *Nigeria Speaks.* London: Longman.

Barkin, David. 1982. "Internationalization of Capital: An Alternative Approach." In Ronald H. Chilcote (ed.), *Dependency and Marxism: Toward a Resolution of the Debate.* Boulder, Colo.: Westview Press.

Barnet, Richard J., and Ronald E. Muller. 1974. *Global Reach: The Power of the Multinational Corporations.* New York: Simon & Schuster.

Bauer, Peter. 1959. "International Economic Developments." *Economic Journal* (March): 112.

Bello, V.I. 1975. "The Intentions, Implementation Process and Problems of the Nigerian Enterprises Promotion Decree (No. 4) 1972." In *Nigeria's Indigenisation Policy. Proceedings of the 1974 Symposium Organised by the Nigerian Economic Society.* Ibadan: The Caxton Press.

Berger, Manfred. 1975. *Industrialisation Policies in Nigeria.* Munich: Weltforun Verlag.

Bienen, Henry, and V.P. Diejomaoh (eds.). 1981. *The Political Economy of Income Distribution in Nigeria.* New York: Holmes & Meier.

Biersteker, T.J. 1978. *Distortion or Development? Contending Perspectives on the Multinational Corporation.* Cambridge, Mass.: MIT Press.

Bohi, Douglas R., and W. David Montgomery. 1982. *Oil Prices, Energy Security, and Import Policy.* Washington, D.C.: Resources for the Future.

Breese, G. 1966. *Urbanization in Newly Developing Countries.* Englewood Cliffs, N.J.: Prentice-Hall.

Brenton, A. 1964. "The Economics of Nationalism." *Journal of Political Economy* 72 (April):376–86.

Brigham, Eugene F. 1979. *Financial Management: Theory and Practice.* Hinsdale, Ill.: Dryden Press.

Brookfield, H. 1975. *Interdependent Development.* Pittsburgh: University of Pittsburgh Press.

Buchanan, K.M., and J.C. Pugh. 1966. *Land and People in Nigeria.* London: University of London Press.

Burns, Sir Alan. 1969. *History of Nigeria.* London: George Allen & Unwin.

Business International. 1979. *Nigeria: Africa's Economic Giant.* Geneva: Business International.

Business Times. 1976. (July 6):5. "New Indigenisation Schedules."

Business Times. 1978 (24 January):24.

Caiden, Naomi, and Aaron Wildavsky. 1974. *Planning and Budgeting in Poor Countries.* New York: John Wiley Sons.

Cairncross, Alex K. 1961. "International Trade and Economic Development." *Economica* (August): 240.

Cardoso, Fernando H. 1972. "Dependency and Development in Latin America." *New Left Review* 74 (July–August):83–95.

Caves, Richard E. 1974. *International Trade, International Investment, and Imperfect Markets.* Princeton, N.J.: International Finance Section, Princeton University.

_____. 1982. *Multinational Enterprise and Economic Analysis.* Cambridge: Cambridge University Press.

Central Bank of Nigeria. *Annual Report and Statement of Accounts.* (Various volumes).

_____. *Economic and Financial Review.* (Various issues).

Cervenka, Zdenek. 1971. *The Nigerian Civil War. 1967–1970.* Frankfurt: Bernard & Graefe Verlag für Wehrwesen.

Chenery, Hollis B. (ed.). 1979. *Structural Change and Development Policy.* New York: Oxford University Press.

Chilcote, Ronald H. 1974. "Dependency: A Critical Synthesis of the Literature." *Latin American Perspectives* 1 (Fall):4–29. Reprinted in J. Abu-Lughod and R. Hay (eds.). 1977. *Third World Urbanization.* Chicago: Maaroufa Press.

_____. 1978. "A Question of Dependency." *Latin American Research Review* 13, no. 2:57–58.

_____. 1980. "Theories of Dependency: The View from the Periphery." In I. Vogeler and A. de Souza (eds.), *Dialectics of Third World Development.* Montclair, N.J.: Allanheld, Osmun & Co.

_____ (ed.). 1982. *Dependency and Marxism: Toward a Resolution of the Debate.* Boulder, Colo.: Westview Press.

Chilcote, Ronald H., and Dale J. Johnson. 1983. *Theories of Development: Mode of Production or Dependency?* Beverly Hills, Calif.: Sage Publications.

Chipman, John S. 1965–66. "A Survey of the Theory of International Trade: Part 1: The Classical Theory." *Econometrica* 33 (October):477–519; "Part 2: The Neoclassical Theory." *Econometrica* 33 (October):685–760; "Part 3: The Modern Theory." *Econometrica* 34 (January):18–76.

Clifford, Hugh. 1918. *German Colonies: A Plea for the Native Races.* London: John Murray.

Cline, William R., and Sidney Weintraub (eds.). 1981. *Economic Stabilization in Developing Countries.* Washington, D.C.: Brookings Institution.

Clower, R.W., et al. 1966. *Growth Without Development.* Evanston, Ill.: Northwestern University Press.

Coleman, James S. 1958. *Nigeria: Background to Nationalism.* Berkeley: University of California Press.

Colonial Office. 1945. *A Ten-Year Plan of Development and Welfare for Nigeria.* Sessional no. 24. Lagos: Government Printer.

Constitution of the Federal Republic of Nigeria. 1978. Apapa, Nigeria: Times Press.

Crowder, Michael. 1973. *The Story of Nigeria.* London: Faber & Faber.

Daily Times. 1979. (January 9):3.

Daly, D.J., and S. Globerman. 1976. *Tariff and Science Policies: Application of a Model of Nationalism.* Toronto: University of Toronto Press.

Damachi, Ukandi G. 1972. *Nigerian Modernization: The Colonial Legacy.* New York: The Third Press.

David, Wilfred L. (ed.). 1973. *Public Finance, Planning and Economic Development.* New York: St. Martin's Press.

Dean, Edwin. 1972. *Plan Implementation in Nigeria: 1962–1968.* Ibadan: Oxford University Press.

Denoon, David B.H. (ed.). 1979. *The New International Economic Order: A U.S. Response.* New York: New York University Press.

Di Marco, Luis E. (ed.). 1972. *International Economics and Development: Essays in Honor of Raul Prebisch.* New York: Academic Press.

Diamond, Larry. 1983. "Social Change and Political Conflict in Nigeria's Second Republic." In I.W. Zartman, (ed.), *The Political Economy of Nigeria*:25–84. New York: Praeger.

Dimka's Confession: The Tragedy of a Nation. Benin, Nigeria: Bendel Newspapers Corp., 1976.

Domar, Evsey D. 1957. *Essays in the Theory of Economic Growth.* New York: Oxford University Press.

Dorrance, G.S. 1948–1949. "The Income Terms of Trade." *Review of Economic Studies*:50–56.

Dos Santos, Theotonio. 1970. "The Structure of Dependence." *American Economic Review* 60 (May):231–36.

_____. 1978. *Imperialismo y dependencia.* Mexico City: Ediciones Era.

Duesenberry, J.S. 1949. *Income, Saving and the Theory of Consumer Behavior.* Cambridge: Cambridge University Press.

Eicher, Carl K., and Carl Liedholm (eds.). 1970. *Growth and Development of the Nigerian Economy.* East Lansing: Michigan State University Press.

Ejiofor, Pita N.O. 1976. "Multinational Corporations as Agents of Imperialism." In B.O. Onibonoje, K. Omotosho, and O.A. Lawal (eds.), *The Indigenes for National Development*. Ibadan: Onibonoje Publishers.

Ekukinam, A.E. 1975. "Opening Address." In *Nigeria's Indigenisation Policy. Proceedings of the 1974 Symposium Organised by the Nigerian Economic Society*. Ibadan: Caxton Press.

Ekundare, R. Olufemi. 1973. *An Economic History of Nigeria*. New York: Africana Publishing Co.

El Mallakh, Ragaei. 1982. *Saudi Arabia: Rush to Development*. Baltimore: Johns Hopkins University Press.

Eleazu, Uma O. 1977. *Federalism and Nation-Building*. Elms Court, Devon: Arthur H. Stockwell.

Emmanuel, Aghiri. 1972a. "White-Settler Colonialism and the Myth of Investment Imperialism." *New Left Review* 73 (May–June):20–34.

_____. 1972b. *Unequal Exchange: A Study of the Imperialism of Trade*. New York: Monthly Review Press.

_____. 1974. "Myths of Development versus Myths of Underdevelopment." *New Left Review* 75 (May–June):61–82.

Federal Republic of Nigeria (FRN). *Annual Abstract of Statistics*. Lagos: Federal Office of Statistics. (Various volumes).

_____. 1964. *National Development Plan: Progress Report*. Lagos: Federal Ministry of Economic Development, March.

_____. 1966. *Guideposts for Second National Development Plan*. Lagos: Ministry of Economic Development, June.

_____. 1968. *Background Notes on the Nigerian Crisis*. Lagos: Federal Ministry of Information.

_____. 1970. *Second National Development Plan, 1970–74. Programme of Post-War Reconstruction and Development*. Lagos: Federal Ministry of Information.

_____. 1972. "Nigerian Enterprises Promotion Decree 1972." In supplement to *Official Gazette Extraordinary* 59, no. 10 (February).

_____. 1973. *Guidelines for the Third National Development Plan 1975–80*. Lagos: Central Planning Office.

_____. n.d. *Second National Development Plan 1970–74. First Progress Report*. Lagos: Central Planning Office.

_____. 1974a. *Second National Development Plan 1970–74. Second Progress Report*. Lagos: Central Planning Office.

_____. 1975a. *Third National Development Plan 1975–1980*. Special Launching Edition. Lagos: Central Planning Office.

_____. 1975b. *First Progress Report on Third National Development Plan 1975–80*. Lagos: Central Planning Office.

_____. 1976a. *Federal Military Government's View on the Report of the Industrial Enterprises Panel*. Lagos: Federal Ministry of Information.

_____. 1976b. *Report of the Constitution Drafting Committee*. Vol. I. Lagos: Constitution Drafting Committee.

_____. 1976c. *Report of the Tribunal of Inquiry into the Importation of Cement*. Lagos: Ministry of Information.

_____. 1979a. *Economic and Statistical Review 1978*. Lagos: Federal Government Press.

_____. 1979b. *Recurrent and Capital Estimates of the Federal Republic of Nigeria, 1979–80*. Lagos: Federal Ministry of Information.

_____. 1984a. *News About Nigeria*. I, no. 7. New York: Nigerian Information Services, February.

_____. 1984b. *An Address to the Nation on the 1984 Budget by Major-General Muhammadu Buhari, C.F.R., Head of the Federal Military Government, Commander-in-Chief of the Armed Forces on 7th May, 1984*. Washington, D.C.: Nigeria Information Service.

_____. 1984c. *Statement on the 1984 Budget by Dr. O.O. Soleye, Minister of Finance*. Special news release. New York: Consulate General of Nigeria, May 9.

Federation of Nigeria. 1959. *Economic Survey of Nigeria*. Lagos: National Economic Council.

_____. 1961. *National Development Plan 1962–68*. Lagos: Federal Ministry of Economic Development.

Fesharaki, Fereidun, and David T. Isaac. 1981. *OPEC Downstream Processing— A New Phase of the World Oil Market*. Honolulu: East-West Center.

Flanders, M.J. 1964. "Prebisch on Protectionism: An Evaluation." *Economic Journal*, LXXIV (June):305–26.

Frank, Andre G. 1966. "The Development of Underdevelopment." *Monthly Review* 28 (September):17–31.

_____. 1967. *Capitalism and Underdevelopment in Latin America: Historical Studies of Chile and Brazil*. New York: Monthly Review Press.

_____. 1975. *On Capitalist Underdevelopment*. Bombay: Oxford University Press.

_____. 1978a. *Dependent Accumulation and Underdevelopment*. London: Macmillan.

_____. 1978b. *World Accumulation: 1492–1789*. London: Macmillan.

_____. 1981. *Reflections on the World Economic Crisis*. New York: Monthly Review Press.

_____. 1983. "The Crisis and Transformation of Dependency in the World-System." In Ronald H. Chilcote and Dale J. Johnson (eds.), *Theories of Development: Mode of Production or Dependency?* Beverly Hills, Calif.: Sage Publications.

Frondizi, Silvio. 1954. *La integracion mundial, ultima etapa del capitalismo (respuesta a una critica) 1947.* Buenos Aires: Praxis.

_____. 1957. *La realidad argentina: Ensayo de interpretacion sociologica.* 2 vols. Buenos Aires: Praxis.

Furtado, Celso. 1963. *The Economic Growth of Brazil: A Survey from Colonial to Modern Times.* Berkeley: University of California Press.

_____. 1964. *Development and Underdevelopment.* Berkeley: University of California Press.

Geary, Sir William N.M. 1965. *Nigeria Under British Rule.* New York: Barnes & Noble.

Ghadar, Fariborz. 1977. *The Evolution of OPEC Strategy.* Lexington, Mass.: Lexington Books/D.C. Heath.

Gill, Richard T. 1963. *Economic Development: Past and Present.* Englewood Cliffs, N.J.: Prentice-Hall.

Gonzalez Casanova, Pablo. 1970. *Democracy in Mexico.* New York: Oxford University Press.

Government Printer. 1946. *A Ten-year Plan of Development and Welfare for Nigeria 1946.* Lagos: Government Printer.

Great Nigeria People's Party. 1979. *The Great People's Charter. Being the Manifesto, Aims and Objectives, and Programme of Action of the Great Nigeria People's Party.* Lagos: GNPP.

Gutkind, Peter C.W., and Immanuel M. Wallerstein. 1976. *The Political Economy of Contemporary Africa.* Beverly Hills, Calif.: Sage Publications.

Haberler, Gottfried. 1958. "International Trade and Economic Development." In James D. Theberge (ed.), *Economics of Trade and Development.* New York: John Wiley & Sons.

Hagen, E.E. 1972. "Economic Growth with Unlimited Foreign Exchange and No Technical Progress." In J. Bhagwati and R.S. Eckaus (eds.), *Development Planning.* London: George Allen & Unwin.

_____. 1975. *The Economics of Development.* Homewood, Ill.: Richard D. Irwin.

Hallwood, Paul, and Stuart Sinclair. 1981. *Oil, Debt, and Develoment: OPEC in the Third World.* London: George Allen & Unwin.

Harbison, Frederick H., et al. 1970. *Quantitative Analyses of Modernization and Development.* Princeton, N.J.: Princeton University Press.

Harrod, Roy F. 1939. "An Essay in Dynamic Theory." *Economic Journal* 49 (March):14–33.

———. 1963. *Towards a Dynamic Economics.* New York: Macmillan.

Hatch, John. 1970. *Nigeria: The Seeds of Disaster.* Chicago: Henry Regnery Co.

Heady, Ferrel. 1979. *Public Administration: A Comparative Perspective.* New York: Marcel Decker.

Heimann, E. 1952. "Marxism and Underdeveloped Countries." *Social Research* (September).

Helleiner, Gerald K. 1970. "The Fiscal Role of the Marketing Boards in Nigerian Economic Development, 1947–61." In C.K. Eicher and Carl Liedholm (eds.), *Growth and Development of the Nigerian Economy.* East Lansing: Michigan State University Press.

Higgins, Benjamin H. 1956. "Development Planning and the Economic Calculus." *Social Research* (Spring):35–56.

———. 1968. *Economic Development: Principles, Problems and Policies.* New York: W.W. Norton.

Hirschmann, Albert O. 1958. *The Strategy of Economic Development.* New Haven: Yale University Press.

Hymer, Stephen. 1972. "The Multinational Corporation and The Law of Uneven Development." In Jagdish N. Bhagwati (ed.), *Economics and World Order: From the 1970s to the 1990s.* London: Macmillan.

Igbozurike, Martin. 1976. *Problem-Generating Structures in Nigeria's Rural Development.* Uppsala: Scandinavian Institute of African Studies.

International Bank for Reconstruction and Development (IBRD). 1955. *The Economic Development of Nigeria. Report of a Mission Organized by the IBRD at the Request of the Governments of Nigeria and the United Kingdom.* Baltimore: Johns Hopkins University Press.

International Monetary Fund. 1984. *International Financial Statistics Yearbook 1984.* Washington, D.C.: IMF.

International Petroleum Encyclopaedia. 1981. Tulsa, Okla.: Penwell Publishing Co.

Jacoby, N.H. 1975. *Multinational Oil.* New York: Macmillan.

Jalee, P. 1968. *The Pillage of the Third World.* New York: Monthly Review Press.

Johnson, Carlos. 1982. "Dependency and Processes of Capitalism and Socialism." In Ronald H. Chilcote (ed.), *Dependency and Marxism: Toward a Resolution of the Debate.* Boulder, Colo.: Westview Press.

Johnson, Harry G. (ed.). 1967a. *Economic Nationalism in Old and New States.* Chicago: University of Chicago Press.

_____. 1967b. *Economic Policies Toward Less Developed Countries.* Washington, D.C.: Brookings Institution.

_____. 1972. "The Ideology of Economic Policy in the New States." In David Wall (ed.), *Chicago Essays in Economic Development.* Chicago: University of Chicago Press.

Keynes, John M. 1933. "National Self-sufficiency." *Yale Review* 22:755–69.

_____. 1936. *The General Theory of Employment, Interest and Money.* New York: Harcourt, Brace.

_____. 1952. *Essays in Persuasion.* London: Rupert Hart-Davis.

Kindleberger, C.P. 1956. *The Terms of Trade: A European Case Study.* New York: Technology Press/Wiley.

Kindleberger, C.P., and B. Herrick. 1977. *Economic Development.* New York: McGraw-Hill.

Kirk-Greene, Anthony, and Douglas Rimmer. 1981. *Nigeria Since 1970: A Political and Economic Outline.* London: Hodder & Stoughton.

Kramer, Ronald K., et al. 1966. *International Trade and Finance: Theory, Policy, Practice.* London: Edward Arnold.

Krause, W. 1965. *International Economics.* Boston: Houghton Mifflin.

Kurihara, Kenneth K. 1959. *The Keynesian Theory of Economic Development.* New York: Columbia University Press.

Kuznets, Simon. 1965. *Economic Growth and Structure: Selected Essays.* New York: W.W. Norton.

_____. 1966. *Modern Economic Growth: Rate, Structure and Spread.* New Haven: Yale University Press.

Lagos Stock Exchange. 1979. *Annual Report and Accounts for the Year Ended 30th September, 1978.* Lagos: Lagos Stock Exchange.

Lenin, V.I. 1967. "Imperialism: The Highest Stage of Capitalism." In *Selected Works.* Vol. I. Moscow: Progress Books.

Levitt, Theodore. 1965. "Exploit the Product Life Cycle." *Harvard Business Review* (November–December):81–94.

Lewis, W. Arthur. 1955. *The Theory of Economic Growth.* Homewood, Ill.: Richard D. Irwin.

_____. 1967. *Reflections on Nigeria's Economic Growth.* Paris: Development Centre of the OECD.

Lipton, M. 1962. "Balanced and Unbalanced Growth in Underdeveloped Countries." *Economic Journal* (September):641–57.

Loehr, William, and John P. Powelson. 1983. *Threat to Development: Pitfalls of the NIEO.* Boulder, Colo.: Westview Press.

Loup, Jacques. 1980. *Can the Third World Survive?* Baltimore: Johns Hopkins University Press.

Mabogunje, Akin L. 1968. *Urbanization in Nigeria.* London: London University Press.

_____. 1971. *Growth Poles and Growth Centers in the Regional Development of Nigeria.* Geneva: United Nations Institute for Social Research.

_____. (ed.). 1973. *Kainji. A Nigerian Man-made Lake.* Ibadan: Institute of Social and Economic Research.

_____. 1974. "Urbanization Problems in Africa." In S. El-Shakhs and R. Obudho (eds.), *Urbanization, National Development, and Regional Planning in Africa.* New York: Praeger.

_____. 1981. *The Development Process: A Spatial Perspective.* New York: Holmes & Meier.

MacPherson, Stewart. 1982. *Social Policy in the Third World: The Social Dilemma of Underdevelopment.* Brighton, Sussex: Wheatsheaf Books.

MacPherson, Stewart. 1982. *Social Policy in the Third World: TRhe Social Dilemma of Underdevelopment.* Brighton, Sussex: Wheatsheaf Books.

Malthus, Thomas R. 1951. *Principles of Political Economy* (1836). New York: Augustus Kelly Reprint.

_____. 1963. *An Essay on the Principle of Population or A View of Its Past and Present Effects on Human Happiness with an Inquiry into Our Prospects Respecting the Future Removal or Mitigation of the Evils Which It Occasions* (1798, 1803). Homewood, Ill.: Richard D. Irwin.

Marcussen, Henrik S., and Jens E. Torp. 1982. *The Internationalization of Capital: Prospects for the Third World.* Uppsala: Scandinavian Institute of African Studies.

Marini, Ruy Mauro. 1969. *Subdesarrollo y revolucion.* Mexico City: Siglo Veintiuno Editores.

_____. 1973. *Dialectica de la dependencia.* Mexico City: Ediciones Era.

Marshall, Alfred. 1919. *Industry and Trade.* London: Macmillan.

_____. 1920. *Principles of Economics.* London: Macmillan.

_____. 1958. *Elements of Economics.* Vol. I. New York: St. Martin's Press.

Marx, Karl. 1946. *Pre-capitalism Economic Formations.* New York: International Publishers.

_____. 1967. *Capital: A Critique of Political Economy.* 3 volumes. New York: International Publishers.

Matthews, R.C.O., C.H. Feinstein, and J.C. Odling-Smee. 1982. *British Economic Growth: 1856–1973.* Stanford, Calif.: Stanford University Press.

McAuslan, Patrick. 1980. *The Ideologies of Planning Law.* New York: Pergamon Press.

Meier, Gerald M. 1968. *The International Economics of Development.* New York: Harper & Row.

_____. 1976. *Leading Issues in Economic Development.* New York: Oxford University Press.

_____. 1977. *Employment, Trade, and Development.* Geneva: Institut Universitaire de Hautes Etudes Internationales.

_____. 1980. *International Economics: The Theory of Policy.* New York: Oxford University Press.

Millor, Manuel R. 1982. *Mexico's Oil: Catalyst for a New Relationship with the U.S.?* Boulder, Colo.: Westview Press.

Morgan, Theodore. 1963. "Trends in Terms of Trade and Their Repercussions on Primary Producers." In R. Harrod and D. Hague (eds.), *International Trade Theory in a Developing World.* London: Macmillan.

_____. 1975. *Economic Development: Concept and Strategy.* New York: Harper & Row.

Morgan, W.T.M. 1984. "Oil Boom out of Steam." *The Geographical Magazine* (March):129–33.

Morning News. 1984. (January 20):A2. "Ex-Nigerian Officials Said to Hold Vast Sums."

Morris, David. 1982. *Self-reliant Cities: Energy and the Transformation of Urban America.* San Francisco: Sierra Club.

Mortimer, Robert A. 1984. *The Third World Coalition in International Politics.* Boulder, Colo.: Westview Press.

Murray, D. 1978. "Export Earnings Instability: Price, Quantity, Supply, Demand?" *Economic Development and Cultural Change* 27 (October):61–73.

Myint, Hla. 1958. "The 'Classical Theory' of International Trade and the Underdeveloped Countries." *Economic Journal* 68 (June):317–37.

_____. 1977. "Adam Smith's Theory of International Trade in the Perspective of Economic Development." *Economica* 44 (August):231–48.

Myrdal, Gunnar. 1956a. *An International Economy: Problems and Prospects.* New York: Harper & Brothers.

_____. 1956b. *Development and Underdevelopment.* Cairo: National Bank of Egypt.

_____. 1957a. *Economic Theory and Underdeveloped Regions.* London: G. Duckworth.

_____. 1957b. *Rich Land and Poor: The Road to World Prosperity.* New York: Harper Books.

_____. 1968. *Asian Drama: An Inquiry into the Poverty of Nations.* London: Penguin Press.

National Economic Council. 1959. *Economic Survey of Nigeria.* Lagos: NEC.

National Party of Nigeria. 1979. "Manifesto of the NPN." In Chuba Okadigbo, ed., *Mission of the NPN.* Enugu: Ejire R. Nwankwo Associates.

New African. 1984. (March):53–70. "Nigeria: What Next?"

New Nigerian. 1977. (5 January). "Alien Firms Ordered to Submit Data."

Nigerian Enterprises Promotion Board. 1977. *Nigerian Enterprises Promotion Decree, 1977.* Lagos: NEPB.

_____. 1979a. *Brief for the Federal Minister of Industries on the Nigerian Enterprises Promotion Decree, 1977.* Lagos: NEPB.

_____. 1979b. *Compliance with the Nigerian Enterprises Promotion Decree, 1977.* Lagos: The Secretariat, NEPB, January 19.

Nigerian Observer 1974. (February 27):1–2. "Scrap Indigenisation Decree . . ."

Nigerian Opinion 1966. (February): "The Last Hurray."

Nigerian People's Party. 1979. *Manifesto, Aims and Objectives, and Programme of Action. 1979.* Lagos: NPP.

Nigerian Yearbook. 1976, 1977–78. Apapa, Nigeria: Times Press.

Niven, Sir Rex. 1967. *Nigeria.* New York: Frederick A. Praeger.

Nkrumah, Kwame. 1963. *Africa Must Unite.* London: Heinemann.

_____. 1964. *Consciencism: Philosophy and Ideology for Decolonisation and Development with Particular Reference to the African Revolution.* London: Heinemann.

_____. 1966. *Neo-colonialism. The Last Stage of Imperialism.* New York: International Publishers.

_____. 1970. *Class Struggle in Africa.* New York: International Publishers.

Noelke, Michael. 1979. *Europe–Third World: The Interdependence File.* Brussels: Commission of the European Communities.

Nurkse, Ragnar. 1953. *Problems of Capital Formation in Underdeveloped Countries.* Oxford: Basil Blackwell.

_____. 1959. *Patterns of Trade and Development.* Wicksell Lectures. Stockholm: Almquist. Reprinted New York: Oxford University Press, 1961.

_____. 1963. "Some International Aspects of the Problem of Economic Development." In Agarwala and Singh (eds.), *The Economics of Underdevelopment:* 256–71. New York: Oxford University Press.

Nwankwo, G.O. 1980. *The Nigerian Financial System*. New York: Africana Publishing Co.

Nweze, C.C. 1980. "Notes on the Nigerian Agricultural Bank." In J.K. Onoh (ed.), *The Foundations of Nigeria's Financial Infrastructure.* London: Croom Helm.

Nyerere, Julius K. 1968. *Ujamaa. Essays on Socialism.* Dar-es-Salaam: Oxford University Press.

_____. 1974. "Our Independence not for Sale." *Africa* no. 33 (May):66–67.

O'Connell, James. 1967. "Political Integration: The Nigerian Case." In A. Hazzlewood, (ed.), *African Integration and Disintegration: Case Studies in Economic and Political Union.* London: Oxford University Press.

Odetola, Theophilus O. 1978. *Military Politics in Nigeria.* New Brunswick, N.J.: Transaction Books.

Ogunsheye, A. 1972. "Experience and Problems of Indigenous Enterprises." In *The Proceedings of the Tenth Annual Conference of the Nigerian Institute of Management: Indigenisation and Economic Development.* Lagos: NIM.

Okafor, F.O. 1980. "The Nigerian Capital Market." In J.K. Onoh (ed.), *The Foundation of Nigeria's Financial Infrastructure.* London: Croom Helm.

Okubadejo, N.A.A. 1969. "Economic Development and Planning in Nigeria 1945–68." In T.M. Yesufu (ed.), *Manpower Problems and Economic Development in Nigeria.* Ibadan: Oxford University Press.

Olaloku, F.A., et al. 1979. *Structure of the Nigerian Economy.* New York: St. Martin's Press.

Olayide, S.O. (ed.). 1976. *Economic Survey of Nigeria, 1960–1975.* Ibadan: Aromolaran Publishing Co.

Olayiwola, P.O. 1985. "Petroleum and Structural Change in a Developing Society: The Case of Nigeria." Ph.D. dissertation, College of Urban Affairs and Public Policy, University of Delaware.

Olorunsola, Victor A. 1977. *Societal Reconstruction in Two African States.* Washington, D.C.: University Press of America.

Oni, O., and B. Onimode. 1975. *Economic Development of Nigeria: The Socialist Alternative.* Ibadan: The Nigerian Academy of Arts, Sciences and Technology.

Onimode, Bade. 1982. *Imperialism and Underdevelopment in Nigeria: The Dialectics of Mass Poverty.* London: Zed Press.

Onoh, J.K. (ed.). 1980. *The Foundations of Nigeria's Financial Infrastructure.* London: Croom Helm.

Onosode, G.O. 1977. "Management Problems of Indigenisation: A Financial View." In U. Udo-Aka et al., *Management Development in Nigeria. The Challenge of Indigenisation.* Ibadan: Oxford University.

Onyemelukwe, C.C. 1974. *Economic Underdevelopment: An Inside View.* London: Longman Group.

OPEC Oil Report 1977. 1977. London: Petroleum Economist, December.

_____ *1979.* 1979. London: Petroleum Economist, November.

Osifo, D.E. 1973. *Indigenisation of Business Enterprises in Nigeria.* Ibadan: Nigerian Institute for Social and Economic Research, University of Ibadan.

Osoba, S.O. 1973. "Ideology and Planning for National Economic Development 1946–72." In M. Tukur and T. Olagunju, eds., *Nigeria in Search of a Viable Polity.* Zaria: Institute of Administration.

Oyebode, A. 1977. "National Participation and the Control of Nigerian Economic Activities." Paper presented at the 15th Annual Conference of the Nigerian Association of Law Teachers, University of Lagos, April 13–16.

Oyediran, Oyeleye, and E. Alex Gboyega. 1979. "Local Government and Administration." In O. Oyediran, (ed.), *Nigerian Government and Politics Under Military Rule, 1966–79.* London: Macmillan Press.

Panter-Brick, Keith (ed.). 1978. *Soldiers and Oil: The Political Transformation of Nigeria.* London: Frank Cass & Co.

Payer, C. 1974. *The Debt Trap: The International Monetary Fund and the Third World.* New York: Monthly Review Press.

Payer, C., and G.K. Shaw. 1971. *The Economic Theory of Fiscal Policy.* New York: St. Martin's Press.

Pearson, Lester B. 1970. *The Crisis of Development.* New York: Praeger.

People's Redemption Party. 1979. *The Platform of the People: The General Programme and Election Manifesto of the Peoples Redemption Party.* Lagos: PRP.

Petersen, William. 1966. "On Some Meaning of Planning." *Journal of the American Institute of Planners* 32 (May):130–42.

Prado Junior, Caio. 1966. *A revolucao brasileira.* Sao Paulo: Editora Brasiliense.

_____. 1967. *The Colonial Background of Modern Brazil.* Suzete Macedo, trans. Berkeley: University of California Press.

Prebisch, Raul. 1950. *The Economic Development of Latin America.* New York: United Nations.

_____. 1959. "Commercial Policy in Underdeveloped Countries." *American Economic Review* 44 (May):251–73.

_____. 1980. "The Dynamics of Peripheral Capitalism." In Louis Lafeber and Liisa L. North (eds.), *Democracy and Development in Latin America.* Studies on the Political Economy, Society and Culture of Latin America and the Caribbean, no. 1, pp. 21–27.

Reedy, Jerry. 1984. *Notable Quotables.* Chicago: World Book Encyclopedia.

Rimlinger, Gaston V., and Carolyn C. Stremlau. 1973. *Indigenisation and Management Development in Nigeria.* Lagos: Nigerian Institute of Management.

Rimmer, Douglas. 1978. "Elements of the Political Economy." In K. Panter-Brick (ed.), *Soldiers and Oil. The Political Transformation of Nigeria.* London: Frank Cass.

_____. 1981. "Development in Nigeria: An Overview." In H. Bienen and V.P. Diejomaoh, eds., *The Political Economy of Income Distribution in Nigeria:* 29–87. New York: Holmes & Meier.

Robertson, Dennis H. 1947. "The Future of International Trade." In Howard S. Ellis and Lloyd Metzler (eds.), *Readings in the Theory of International Trade.* Philadelphia: Richard D. Irwin.

Roll, E. 1965. *A History of Economic Thought.* 3rd ed. New York: Prentice-Hall.

Rosenstein-Rodan, P.N. 1957. *Notes on the Theory of the "Big Push."* Cambridge, Mass.: MIT/CIS.

_____. 1963. "Problems of Industrialization of Eastern and South-Eastern Europe." In A.N. Agarwala and S.P. Singh (eds.), *The Economics of Underdevelopment.* New York: Oxford University Press.

Rosenthal, Robert J. 1984. "How Nigeria Missed Its Big Chance." *Philadelphia Inquirer Magazine* (December 30):1, 11–19, 26, 31.

Rosovsky, Henry. 1965. "The Take-off into Sustained Controversy." *Journal of Economic History* (March):271–75.

Rostow, Walt W. 1950. "The Terms of Trade in Theory and Practice." *Economic History Review* 2nd ser., 1:1–15.

_____. 1962. *The Process of Economic Growth.* New York: Norton.

_____. 1963. "The Take-off into Self-sustained Growth." In A.N. Agarwala and S.P. Singh (eds.), *The Economics of Underdevelopment.* New York: Oxford University Press.

_____. 1964. *The Stages of Economic Growth: A non-Communist Manifesto.* Cambridge, Mass.: Harvard University Press.

Sada, P.O. 1981. "Urbanization and Income Distribution in Nigeria." In H. Bienen and V.P. Diejomaoh (eds.), *The Political Economy of Income Distribution in Nigeria.* New York: Homes & Meier: pp. 29–87.

Salazar-Carrillo, Jorge. 1976. *Oil in the Economic Development of Venezuela.* New York: Praeger.

Sathyamurthy, T.V. 1983. *Nationalism in the Contemporary World. Political and Sociological Perspectives.* London: Frances Pinter.

Sauvant, Karl P., and Hajo Hasenpflug (eds.). 1977. *The New International Economic Order: Confrontation or Cooperation Between North and South?* Boulder, Colo.: Westview Press.

Schatz, Sayre P. 1963. "Nigeria's First National Development Plan (1962–68): An Appraisal." *Nigerian Journal* 5, no. 2:221–35.

Schiavo-Campo, Salvatore, and Hans W. Singer. 1970. *Perspectives of Economic Development.* Boston: Houghton Mifflin.

Schumpeter, Joseph A. 1934. *The Theory of Economic Development.* Cambridge, Mass: Harvard University Press.

_____. 1942. *Capitalism, Socialism and Democracy.* New York: Harper & Row.

_____. 1954. *History of Economic Analysis.* New York: Oxford University Press.

Schwarz, Walter. 1968. *Nigeria.* New York: Frederick A. Praeger.

Scitovsky, Tibor. 1954. "Two Concepts of External Economies." *Journal of Political Economy* 62 (April):143–51.

Scott, William G., and David K. Hart. 1979. *Organizational America.* Boston: Houghton Mifflin.

Seers, Dudley. 1977. "The New Meaning of Development." *International Development Review* no. 3:3.

Shackle, G.L.S. 1983. *The Years of High Theory.* Cambridge: Cambridge University Press.

Sigmund, P.E. (ed.). 1972. *The Ideologies of the Developing Nations.* New York: Praeger.

Singer, Hans W. 1949. "Economic Progress in Underdeveloped Countries." *Social Research* (March).

_____. 1950. "The Distribution of Gains Between Investing and Borrowing Countries." Papers and Proceedings of the Sixty-second Annual Meeting of the American Economic Association, 1949. *American Economic Review* 40 (May):473–85.

_____. 1958. "The Concept of Balanced Growth and Economic Development: Theory and Facts." *Proceedings of the University of Texas Conference on Economic Development.*

Smith, Adam. 1937. *An Inquiry into the Nature and Causes of the Wealth of Nations.* (1776). New York: Modern Library.

South 1984a. (January):23–24. "Nigeria: All Set for a Rough Ride."

_____. 1984b. (February):19–20. "Rise and Fall of the Lootocracy."

_____. 1984c. (July):31–40. "Debt: The IMF's African Nightmare."

Stolper, Wolfgang F. 1966. *Planning Without Facts: Lessons from Nigeria's Development.* Cambridge, Mass.: Harvard University Press.

_____. 1970. "Social Factors in Economic Planning with Special Reference to Nigeria." In C.K. Eicher and C. Liedholm (eds.), *Growth and Development of the Nigerian Economy:* 221–39. East Lansing: Michigan State University Press.

Sunkel, Osvaldo. 1972. "Big Business and Dependencia." *Foreign Affairs* 50 (April):517–31.

Tanzer, Michael. 1969. *The Political Economy of International Oil and the Underdeveloped Countries.* Boston: Beacon Press.

Theberge, James D. (ed.). 1968. *Economics of Trade and Development.* New York: John Wiley & Son.

Thirlwall, A.P. 1976. *Financing Economic Development.* London: Macmillan.

TIME 1982. (April 12):62. "Drowning in Unsold Oil: The Dangerous Collapse of Nigeria's Petroleum Prosperity."

_____. 1983. (April 18):59. "Nigeria: A Vigorous but Fragile Democracy."

_____. 1984a. (January 16):24–25. "The Light That Failed: A Military Coup Brings an Abrupt End to Nigeria's Democratic Experiment."

_____. 1984b. (July 16):43. "The Man in the Diplomatic Crate: A Nigerian Exile's Bizarre Kidnaping Infuriates Thatcher."

_____. 1984c. (July 23):52. "Rooting out Corruption: Offering No Apologies, a New Leader Presses His War Against Indiscipline."

Tims, Wouter. 1974. *Nigeria: Options for Long-Term Development.* Baltimore: Johns Hopkins University Press.

Tinbergen, J. 1967. *Development Planning.* New York: McGraw-Hill.

Todaro, Michael P. 1971. *Development Planning: Models and Methods.* London: Oxford University Press.

_____. 1977. *Economic Development in the Third World.* New York: Longman Group.

Tomori, S., and F.O. Fajana. 1979. "Development Planning." In F.A. Olaloku et al., *Structure of the Nigerian Economy:* 131–46. New York: St. Martin's Press.

Touré, Sékou. 1972. "African Emancipation." In P.E. Sigmund (ed.), *The Ideologies of the Developing Nations:* 226–40. New York: Praeger.

Tsuru, S. 1954. "Keynes Versus Marx: The Methodology of Aggregates." In K. Kurihara (ed.), *Post-Keynesian Economics.* New Brunswick, N.J. and London: Rutgers University Press/Allen & Unwin.

Tukur, M., and T. Olagunju (eds.). 1973. *Nigeria in Search of a Viable Polity.* Zaria: Institute of Administration.

Turner, T. 1976. "Multinational Corporations and the Instability of the Nigerian State." *Review of African Political Economy* 5 (January):7–19.

Ukpong, Ignatius I. 1979. "Social and Economic Infrastructure." In F.A. Olaloku, et al. (eds.). *Structure of the Nigerian Economy:* 68–99. New York: St. Martin's Press.

U.S. News & World Report. 1983. (24 October):30–31. "Is OPEC Going the Way of All Cartels?"

United Bank for Africa. 1984. "Seminar on 'Foreign Debt and Nigeria's Economic Development': Summary of Recommendations." *West Africa* (June 4):1176–78.

Unity Party of Nigeria. 1979. *Up Nigeria: Manifesto of the Unity Party of Nigeria.* Lagos: UPN.

Uzamere, A. Osaheni. 1974. "What Happened to Our Business Take-over." *Nigerian Herald* (24 April):5.

Valdes, A. (ed.). 1980. *Food Security for Developing Countries.* Washington, D.C.: IFPRI.

Vallenilla, Luis. 1975. *Oil: The Making of a New Economic Order.* New York: McGraw-Hill.

Viner, Jacob. 1937. *Studies in the Theory of International Trade.* New York: Harper.

Vogeler, Ingolf, and Anthony R. de Souza (eds.). 1980. *Dialectics of Third World Development.* Montclair, N.J.: Allanheld, Osmun & Co.

Wall, David (ed.). 1972. *Chicago Essays in Economic Development.* Chicago: University of Chicago Press.

Wall Street Journal 1982. (July 7). "In Nigeria, Payoffs Are a Way of Life."

_____. 1984. (February 1):1, 17. "Unruly Land: Military Ruler Fights to Discipline Nigeria and End Corruption."

Wallerstein, Immanuel. 1974a. "Dependence in an Interdependent World: The Limited Possibilities of Transformation Within the Capitalist World Economy." *African Studies Review* 17 (April):1–25.

_____. 1974b. *The Modern World-System: Capitalist Agriculture and the Origins of the European World-Economy in the Sixteenth Century.* New York: Academic Press.

_____. 1974c. "The Rise and Future Demise of the World Capitalist System: Concepts for Comparative Analysis." *Comparative Studies in History and Society* 16 (September):387–415.

_____. 1976. "Semi-peripheral Countries and the Contemporary World Crisis." *Theory and Society* 3:461–83.

_____. 1978. "World-System Analysis: Theoretical and Interpretative Issues." In Barbara H. Kaplan (ed.), *Social Change in the Capitalist World Economy:* 219–35. Beverly Hills, Calif.: Sage Publications.

_____. 1979. *The Capitalist World-economy: Essays.* Cambridge: Cambridge University Press.

_____. 1983. "Crises: The World-Economy, the Movements, and the Ideologies." In Albert Bergesen (ed.), *Crises in the World System:* 21–36. Beverly Hills, Calif.: Sage Publications.

_____. 1984. *The Politics of the World-economy: The States, the Movements, and the Civilizations: Essays.* Cambridge: Cambridge University Press.

West Africa. 1982. (February 8):377–88. "History Is Made: Nigeria Produces Steel."

_____. 1983a. (January 24):191–92. "Fruitless Probes."

_____. 1983b. (October 24):2480. "Shehu's Plea to Legislators."

_____. 1983c. (October 31):2538. "49 Factories Closed."

_____. 1984a. (January 9):53–58. "The Return of the Military."

_____. 1984b. (February 27):423. "OPEC and Nigeria."

_____. 1984c. (February 27):424–27. "Interview with General Buhari."

_____. 1984d. (February 27):473. "How Money Was 'Siphoned Abroad.' "

_____. 1984e. (April 2):718–19. "Military Rule in Nigeria: 2; Consistency in Economics."

_____. 1984f. (April 9):793. "War Against Indiscipline."

_____. 1984g. (April 16):815–16. "Buhari's First One Hundred Days."

_____. 1984h. (July 9):1394–95. "The Return of the Nigerian Military: 1; The Pattern of Military Rule."

_____. 1984i. (July 23):1476–77. "Nigeria: 'We Can Sell the Extra Oil.' "

_____. 1984j. (July 23):1477. "Dikko—the Froth Dies Down."

_____. 1984k. (September 3):1817. "Buhari on His Priorities."

Wilcox, Clair. 1965. *The Planning and Execution of Economic Development in Southeast Asia.* Cambridge, Mass: Harvard University Center for International Affairs.

World Bank. 1976. *World Tables, 1976.* Washington, D.C.: World Bank.

_____. 1980. *World Tables, 2nd Edition (1980).* Washington, D.C.: World Bank.

_____. 1982. *World Bank Atlas, 1981.* Washington, D.C.: World Bank.

_____. 1983a. *World Development Report 1983.* New York: Oxford University Press.

_____. 1983b. *World Tables. 3rd Edition (1983).* 2 vols. Baltimore: Johns Hopkins University Press.

Zartman, I. William (ed.). 1983. *The Political Economy of Nigeria.* New York: Praeger.

Zartman, I. William, and Sayre Schatz. 1983. "Introduction." In I.W. Zartman (ed.), *The Political Economy of Nigeria:* 1–24. New York: Praeger.

Index

About the Author

PETER O. OLAYIWOLA is a lecturer in the Department of Government and Public Administration, University of Ilorin, Nigeria.

Dr. Olayiwola holds a B.B.A. and an M.Admin. from the Pennsylvania State University and a Ph.D. from the University of Delaware, Newark. He is a member of the British Institute of Management, the Urban Affairs Association, the Nigerian Economic Society and the Nigerian Political Science Association.

For his original work, Dr. Olayiwola received the Mark A. Haskell Award in Political Economy at the University of Delaware in May 1985.